Advances in Business Ethics Research

A Journal of Business Ethics Book Series

Volume 7

More information about this series at http://www.springer.com/series/8805

Gilbert Lenssen • Jay Hyuk Rhee
Fabien Martinez
Editors

The Role of Corporate Sustainability in Asian Development

A Case Study Handbook in the Automotive and ICT Industries

Springer

ABIS
The Academy of Business in Society

Editors
Gilbert Lenssen
ABIS - The Academy of Business
in Society
Ixelles, Brussels, Belgium

Jay Hyuk Rhee
International Business/Strategy
Korea University Business School
Seoul, South Korea

Fabien Martinez
Uccle, Brussels Hoofdst.ge., Belgium

Advances in Business Ethics Research
ISBN 978-3-319-45158-9 ISBN 978-3-319-45160-2 (eBook)
DOI 10.1007/978-3-319-45160-2

Library of Congress Control Number: 2016952863

Printed on acid-free paper

This Springer imprint is published by Springer Nature
The registered company is Springer International Publishing AG
The registered company address is: Gewerbestrasse 11, 6330 Cham, Switzerland

Preface

Corporate social responsibility (CSR) has been attracting increasing attention from the popular and business media. Popular business magazines including *Business Week* and *Economist* have produced a significant number of stories and articles related to CSR. Such public focus on CSR poses tremendous challenges especially for multinational corporations. Being very visible targets for policy-makers and nongovernmental organizations, multinational companies are under great pressure to become more socially responsible.

The extent of being socially responsible may significantly vary along a continuum ranging from the most reactive postures to the most proactive ones. As such popular terms as "pollution havens" and a "race to the bottom" indicate, "doing well by doing good" on a global basis is not as easy as one might expect. From the multinational companies' point of view, the real challenge is how to capitalize on their CSR practices.

Although each multinational company can internally conduct its own CSR evaluation, an external evaluation based on common criteria and indicators may result in more objective results. The effect of CSR activities is often beyond the companies' immediate financial performance, which underlines the importance of an objective measure and evaluation method to identify corporate CSR efforts. Based on such external evaluation results, multinational companies can understand their current status compared to their peers. External evaluation results also enable stakeholders to identify the extent to which multinational companies are socially responsible not only in their home countries but also in their diverse host countries including emerging or underdeveloped ones.

Acknowledging the need for conducting an external CSR evaluation based on common criteria and indicators, I have been conducting CSR rankings with InnoCSR. Since 2011, we have conducted the Fortune China CSR Ranking with Fortune China every year. The Fortune China CSR Ranking is a leading CSR ranking in China, which takes into account all relevant CSR aspects of Environmental, Social, and Governance (ESG) for companies operating in China. The framework is benchmarked against globally recognized ISO 26000. From this year, we expand our CSR ranking coverage from China to other Asian countries including South Korea, Japan, and ASEAN countries.

CSR has a short history in Asia, but it has been getting more attention recently. Since corporate sustainability can be defined as a business approach that responds to multiple stakeholders on every dimension of how a business operates and creates long-term shared value through integration of business strategies, human values, and ecological culture, it seems logical to think that a fast-growing economy of Asia is boosted by the commitments of businesses in the area for sustainable development.

Recently, more companies in Asia have started to recognize the strategic importance of building practices that create sustainable bottom lines related to the economy, environment, and society. Understanding CSR as a part of obligations of a company, which extends beyond economic and legal obligations, multinational companies have tried to meet their ethical responsibilities by embodying standards and norms that reflect the concern of various stakeholders from consumers and employees to the whole community. One of the representative trends regarding ethical responsibilities in CSR is "Fair Trade" in South Korea. Ethical goals have been also found to be critical to the maintenance of a healthy society and to the need and welfare of different stakeholders.

Other CSR activities commonly practiced in Asia are related to philanthropic responsibilities, usually related to activities that enhance human welfare and goodwill in the respective regions. Most companies that carry out philanthropic projects make donations for such purposes as children or job educations, improvements in community infrastructure, and developments in art and culture. Eventually, CSR activities of multinational companies have been positively affecting the communities by satisfying various stakeholders and developing social welfare as a whole. At the same time, the companies themselves have seen positive effects in the long term regarding brand image and customer satisfaction as well as financial performance.

CSR is no longer the preserve of moralists but rather a common interest of practitioners and scholars. As Director of the Asian Business Center at Korea University Business School, I organized two CSR conferences in Korea in 2012 and 2013 to open up new opportunities and insights. Cohosted with the Academy of Business in Society, both conferences brought interested academics together to share and discuss research on the role of corporate sustainability in the development of the region.

The first conference in 2012 focused on the theme "Corporate Social Responsibility (CSR) and Globalization" with three sessions: (1) CSR as a Global Strategy, (2) CSR in Practice, and (3) How to Assess and Improve CSR. Through this conference, researchers established their relationships with CSR experts and deepened their potential commitment to the professional development in the field of CSR. This conference led practitioners and professors to discuss theoretical backgrounds of CSR.

The second conference in 2013 invited practitioners from the electronics and the automotive industries to discuss "The Role of Corporate Sustainability in Asia's development," focusing more on the practical application of CSR. In particular, global automotive companies such as Hyundai Motor Company from Korea, BMW from Germany, and Mahindra & Mahindra from India had participated in the conference; in the electronics industry, Samsung Electronics from Korea, Intel from the United States, and Lenovo and ZTE from China participated. The conference featured presentations on the sustainability cases of multinational companies that have contributed to the development of Asia and set up an open platform to discuss advances and diverse approaches in the challenged areas.

This book is the result of the discussions from the two conferences regarding current challenges in CSR and the best CSR practices that have contributed to the development of Asia or that can further be applied in order to expand their positive impact in Asia. We have recognized the need for publishing these cases for the purpose of sharing a range of best practices that have been highlighted as relevant in recent years and are certainly salient to address the role of corporate sustainability in Asian development. We hope that this book will be useful to academicians who teach and research CSR issues, practitioners who are searching for appropriate CSR strategies to benchmark, students who are studying to be future business leaders in the field, and the general public who is interested in the CSR activities of multinational companies.

We would like to express our gratitude to all academic researchers and company representatives who contributed to this volume. The financial support for the two conferences by the Institute for Business Research and Education at Korea University Business School, the Kookmin Bank, and the Hyundai Motor Group is also gratefully acknowledged. We would also like to thank Sam Lee, Group CEO of InnoCSR, for his support. Lastly, we were very lucky to count on the support of Elena Urizar and Dr. Ludwig Roger (ABIS) and Jeena Park (InnoCSR) whose consistent efforts of coordination were crucial in the development of this volume.

Seoul, South Korea — Jay Hyuk Rhee
Brussels, Belgium — Gilbert Lenssen
London, United Kingdom — Fabien Martinez

Contents

Chapter 1
Toward Inclusive Economic, Social and Environmental Progress in Asia: An Introduction

Fabien Martinez

The Economic Development and Sustainability Nexus in Asia

The economic development of Asian countries has consistently received priority attention in the region's development policies, programmes and investment strategies of the past decades (Welford 2005). In the 1980s, Asia entered a period of considerable economic growth and technological progress, in tandem with openness to trade within the world economy (Rock et al. 1999). Meanwhile, and perhaps as a consequence, Asian corporations and entrepreneurs started to leverage resources and capabilities to enact various forms of pro-sustainability practices, often focused on localised socio-ecological issues and based on cultural traditions at a country level (Smith and Jalal 2000).

Despite the progress made locally, there is no shortage of evidence-based reports and scholarly works indicating a downward spiral of worsening social and environmental conditions in the region (ADB 2012; Bauer and Thant 2010; Laruelle and Peyrouse 2012; Lipovsky 1995). Economic growth, and by extension the sustainable development of Asian societies, are threatened by two alarming socio-economic trends: rising income inequality and ecological deterioration (Petri and Vinod 2013). A report from the Asian Development Bank (ADB 2012) indicates that rising income inequality is widespread in Asia. It affects eleven countries and covers eighty-two per cent of the region's population. The Gini coefficient, a measure of inequality,[1] rose about 1.4% a year between the mid-1990s and the late 2000s, resulting in a large mass of 'excluded' people. The second trend relates to the deterioration of environmental conditions in Asia. The ADB (2012) throws light on the

[1] In the World Development Report 2006, Gini coefficients are calculated based on disposable income data from household surveys.

F. Martinez (✉)
Queen Mary University of London, School of Business and Management, London, UK
e-mail: f.martinez@qmul.ac.uk

G. Lenssen et al. (eds.), *The Role of Corporate Sustainability in Asian Development*, Advances in Business Ethics Research 7,
DOI 10.1007/978-3-319-45160-2_1

severe ecological damage, such as loss of biodiversity, that has either occurred or has narrowly been averted. In 2013, developing Asia accounted for thirty-five per cent of global dioxide emissions – a figure that was projected to rise to forty-four per cent by 2030. Much of this is due to China and India, the largest and third-largest emitters of carbon dioxide in the world respectively. If these growing emissions continue to be left unchecked, they could raise global temperatures by four degree Celsius by the end of the XXI century (The World Bank 2012).

Asian economic actors are increasingly aware about their risky exposure to future climate change and impacts of ongoing social deprivation. Companies have participated in the deterioration of these socio-ecological issues, and their activities are in turn increasingly affected. The question of how they might act to avert, and perhaps reverse, these negative trends by combining economic progress with strong social and environmental responsiveness has attracted a high level interest among international bodies (e.g., OECD, WBCSD, WWF), academic scholars, management practitioners and policy-makers seeking to develop an adapted and comprehensive sustainability framework for managing environmental resources and social capitals in this region. The number of Asian companies using the Global Reporting Initiative to report on Corporate Social Responsibility (CSR) has grown rapidly in recent years. Mandatory requirements for CSR reporting have been developed in Indonesia. In some countries such as Malaysia and the Philippines, national stock exchanges have adjusted their listing requirements to encourage greater corporate transparency on environmental and social issues. What is more, a number of books and journal articles have been written on this topic and the practices and behaviours of Asian business organisations have consistently emerged as an important research area to explore, either for cross-country and cross-continent comparative analysis (e.g., Matten and Moon 2008; Tipton 2009; Welford 2005; Whitley 1992; C. Williams and Aguilera 2008) or for the study of country-specific corporate sustainability practices (e.g., Fukikawa and Moon 2004; Moon and Shen 2010; Redding and Witt 2009; Wong 2009).

If, to some extent, a firm's capacity to respond to sustainability issues is positively linked to the economic development of the country in which it operates, the strong priority on growth meant that Asian economies typically kept social and environmental expenditures relatively low and focused investments on economic infrastructure. The non-economic dimensions of development, related to individual and communal wellbeing and ecological integrity, have remained an appendage of corporate strategies and operations, severely marginalised, under-resourced and overly dependent on governments' intervention. The difficulties in delivering CSR above and beyond economic performance objectives can be explained by a combination of contingent factors. One important factor is the level of corruption in East Asia, epitomised by the existence of stable and mutually beneficial exchanges of government privileges for bribes and kickbacks (Rock and Bonnett 2004). Another factor that can be highlighted is the financial crisis that emerged in many Asian countries in the late 1990s – arguably (and partly) initiated by irresponsible acts of big corporations borrowing money beyond their returning capacity in the name of rapid expansion (Almunawar and Low 2014; Corsetti et al. 1998; Mishkin 1999).

As a result, the capacity of many Asian economic actors to contribute meaningfully to the search for a more balanced approach to industrial growth, social welfare and ecological integrity was severely compromised (Bauer and Thant 2010; Chaponnière 2012). But the emergence of the crisis, and its public visibility, also provided an opportunity to rethink the established model of development in the region and make growth more inclusive and sustainable. Research began to focus on Asia's vaunted methods of learning and innovation to meet the new challenges instead of advocating for the exclusive adoption of North American and Western European concepts and tools that are generally not acculturated to Asian environment and values (Almunawar and Low 2014). However, our understanding of corporate sustainability-oriented practices in Asia remains largely underdeveloped, mainly due to a lack of practical insights into this phenomenon (Zhao 2014). The collection of illustrative company cases that is offered in this volume, and that focuses on sustainability-oriented enterprises and innovations in Asia, is thus timely. It contributes to the development of a better and more context-sensitive way of integrating concerns for social and environmental impacts into management decisions.

Perspectives on Sustainability in Asian Business: Toward a Distinctively 'More' Integrative Approach

The salient role of business is particularly evident in the literature covering the economic, social and environmental challenges of sustainable development in Asia (e.g., Fukushi et al. 2010; Gillet al. 2010; Lin-Heng et al. 2010; OECD 2001), to which this book constitutes a valuable additional resource. It is more specifically addressed in studies on business sustainability and/or CSR in Asia (e.g., Asia Monitor Resource Center 2012; Carew-Reid 2009; Chapple and Moon 2005; Kotler et al. 2008; Low et al. 2013; Ong 2008; Rowley et al. 2012; G. Williams 2011). The present volume makes an original contribution by bridging these two important streams of literature insofar as it examines the link between the 'pro-sustainability' (otherwise referred to as CSR) activities of seven multi-national corporations (Intel, Lenovo, Samsung Electronics, ZTE, BMW, Hyundai Motor, Mahindra & Mahindra) and the sustainable development of Asia. A specific emphasis is placed on concepts and ideas that are arguably well suited to support this integrative approach, that is: 'social innovation', 'corporate philanthropy', 'green technology', 'sustainable mobility', 'social enterprise', and 'sustainability-related knowledge building'. In doing so, the book seeks to play a substantial part in expanding the positive impact of these 'pro-sustainability' practices in Asia, where both the rapid growth of populations and the development of agriculture and industry are posing a serious problem on economic sustainability.

Related to our attempt at highlighting business practices that have (and are likely to continue to have) a positive impact in Asia is the literature dealing more broadly with the strategic, political, institutional and socio-ecological challenges related to sustainability in Asia. For example, the OECD's (2001) book provides an interesting perspective from which to understand the importance and complexity of uncovering the (hidden) connections between social, economic and environmental processes in the context of Asia. It essentially explores the role of social protection systems in creating more inclusive societies, thereby offering an insightful social/individual approach to the problem of sustainable development in Asia. Its focus on the role of institutions (particularly those attending to the social needs of individuals) lead the authors to allude to different types of socially harmful activities in which businesses are implicated (e.g. distressful work conditions, anti-competitive behaviours). The OECD's book is however relatively light on explanations of how OECD countries are active (and/or challenged) in promoting the diffusion of socially-responsible business in the name of sustainable development. The work of the OECD may be seen to reflect a recurring observation made in recent research works addressing the issue of sustainable development in Asia. That is, multinational companies have the resources and capabilities to foster the technological, institutional, organisational, social and cultural changes that Asian countries are bound to operate to venture toward a more sustainable and inclusive society (Fukushi et al. 2010; Gill et al. 2010; Lin-Heng et al. 2010). Therefore, future research efforts in Asian sustainable development ought to examine the best practices that will help multi-national companies to foster such changes and expand the scope of their positive societal impact.

A popular perspective from which the role of multi-national companies in Asian sustainable development has often been analysed is CSR (Asia Monitor Resource Center 2012; Low et al. 2013; Williams 2011). For example, Low et al. (2013) provide a comprehensive overview of the practice of CSR in Asia. They observe that CSR is often seen through the lenses of Western thinkers. The application of CSR in Asia, they argue, would benefit from the incorporation of Asian philosophies and thoughts. The book proposes to compare Western and Asian perspectives on CSR and present them in the light of Asian philosophies and thoughts, such as Confucian, Islamic (Koranic), Indian (Vedantic) and other Asian ways of looking at CSR in their own rights and perspectives. In a similar vein, the book edited by G. Williams (2011) examines the diverse meanings generally attached to the concept of CSR in Asia. It essentially highlights the role of Asian companies in leading their own way in markets which are hugely complex and dynamic. The focus is on the factors operating inside the company and influencing perceptions of CSR. The books published by Low et al. (2013) and G. Williams (2011) contribute to our understanding of the perceptions and meanings of CSR in Asia. By contrast, the contents of the present volume will be oriented toward understanding the impact of pro-sustainability (or CSR) activities in Asia. The series of case studies is expected to cover a wide range of challenges related to both internal and external corporate processes and that appear to foster sustainable development in Asia. The focus on 'impact' will help to clarify, and potentially expand, the socio-economic contribution of the business community to Asian development.

One general criticism that can be levelled at the contributions that use the concept of CSR is that they are often undertaken in isolation to the economic, social and environmental contexts of the firm. Despite the substantial challenges faced by Asian companies in managing and fostering sustainable development, it may be advanced that our book will be unique in its ambition to make an explicit link between corporate sustainability practices and the sustainable development of Asia. Both the 'marketing' perspective of Kotler et al. (2008) and the 'leadership/relational' approach adopted by Rowley et al. (2012) constitute an effort in this sense as they make a link between sustainable business performance and the strategic (and decision-making) context of the firm. Yet their focus on the organisational (or 'firm-level') factors that appear to be associated with strong corporate sustainability means that the broader implications for the sustainable development of Asia are still neglected – perhaps owing to a lack of depth and contextualisation of the findings. The difficulty in obtaining a complete overview of the sustainability challenges at the scale of both corporations and the Asian continent might be advanced to explain the lack of context-sensitive insights in existing research. This book plays a part in addressing this gap, partially because its contents stem from discussions between academic scholars and management practitioners. Hence, the impact of disciplinary bias is reduced and the transdisciplinary and integrative scope that the book seeks to generate is facilitated.

Book Overview – Seven Illustrative Company Cases

This volume is a cross-disciplinary attempt to clarify some conceptual and practical issues that surface the societal role of business firms in Asia. It brings together highly diverse perspectives that contribute to our understanding of various forms of corporate engagement in pro-sustainability activities and provide great scope for applications to other company cases in Asian countries. One important pattern that is often advanced to motivate corporate engagement in pro-sustainability activities is the extent to which domestic companies engage in international trade, even where that trade is with other Asian nations (Chapple and Moon 2005). This means that some level of homogeneity in engagement modes are likely to be found across Asian companies operating in countries that are well integrated in the international economy – for example: Malaysia, the Philippines, Singapore and Thailand (Chapple et al. 2014). By contrast, the impact of, and on, pre-existing levels of economic development is not seen as a primary source of motivation for societal engagement by Asian companies, resulting in some level of opacity and a lack of knowledge about the extent to which corporate activities generate inclusive economic, social and environmental progress in Asia. This book guides readers through the story of how seven multi-national companies act to promote, develop and implement specific modes of sustainability-driven enterprises and innovations in China, Korea and India.

The authors develop a narrative of how a particular set of challenges arose for the company, how executives responded to these challenges, with what effects (both desired and undesired), externally and internally and with what impact on a various set of stakeholders (e.g., investors, customers, suppliers, employees, non-governmental organisations). Each case ends with an outlook of future challenges. Our understanding of the context in which these multi-national companies operate is enriched by market and financial information. The cases are organised in seven chapters as follows.

Chapter 2 discusses the evolution of the concept of CSR in China with an emphasis on the innovative practices that Information and Communication Technology corporation Intel put in place to create sustainable shared business and social value. Igel, Yeung and Prince introduce the concept of CSR 3.0, referring to it as an advanced version of CSR that fosters social innovation and strengthens stakeholder relationships. The case of Intel allows to unveil important factors within CSR 3.0 that act to catalyse social innovation, such as cross-pollination of idea systems amongst stakeholders and the coalescing of stakeholders' resources.

Chapter 3 explores the venture philanthropy programme launched by PC makers Lenovo in 2007. It provides yet another angle from which to examine the channels through which CSR strategies can be developed in the context of China. The chapter explains how Lenovo leveraged a set of resources and capabilities to support and empower sustainability-oriented enterprises and not-for-profit organisations to make a significant social contribution.

Chapter 4 examines the concept of corporate social responsibility in relation to the activities of Samsung Electronics, with a particular emphasis on the company's green management initiative that started in Korea and went on to impact the entire memory chip industry, shifting it towards increasing consideration and adoption of green innovations.

Chapter 5 discusses the efforts of ZTE to eliminate digital chasm in Africa. The company deploys green innovation and technologies to improve urban livelihood, build a healthier economic and industrial structure, and a stable and harmonious city while promoting sustainable urban development.

Chapter 6 documents and examines how BMW Group Korea has fostered progress in sustainable mobility through the BMW i project. The BMW i project was initiated by the BMW Group in Germany. It constitutes an innovation to build an electric city car designed for urban use and to ensure that the materials, production process, supply chain, and recycling all adhere to sustainability-oriented principles. Considering the interest in electric cars on the market, BMW's efforts and milestones are reviewed to determine the success of launching environmental friendly individual mobility.

Chapter 7 explains how Hyunday Motor Company supports social enterprises via cooperative relations between its corporate sociocultural team and external expert groups, and by fostering representative, job-creating enterprises.

Chapter 8 provides an overview of Mahindra & Mahindra's initial steps to understand the potential for sustainability and the challenges it could encounter when embarking on a sustainability path. The initial approach of the company, and its supporting management structure, are introduced. An overview of its activities intended to raise general awareness and creating management buy-in will be provided and its Road Map to sustainability explicated. The case study concludes with an overview of achievements and outcomes to date and a summary of it external collaborations and partnerships.

These company cases will arguably appeal to the interest of industry practitioners who seek new opportunities and innovative solutions to substantially foster their contribution to the sustainable development of Asian societies. It will also appeal to international bodies who acknowledge the fundamental social and environmental role of business organisations and recognise the urgency of generating solutions to the issues of environmental integrity/resilience and social equity at a global scale (OECD 2001, 2012; WBCSD 2013; WWF 2012). The challenges of managing and integrating economic, environmental and social responsibility in business also attract a high level interest among policy review agencies (or 'think tanks'), policy-makers and welfare economists seeking to support the development of a comprehensive sustainability framework for Asian development. The volume is anticipated to have an educational objective. It can be used in graduate classrooms to teach and provoke stimulating debates about the theoretical and practical challenges of corporate sustainability in the context of Asia. We further expect that the book will be of interest to a broad range of researchers. It offers a rich source of case study examples that guide toward the systematic development of 'pro-sustainability' corporate initiatives in Asia.

References

ADB. 2012. *Asian development outlook 2012. Confronting rising inequality in Asia*. Mandaluyong City: Asian Development Bank.

Almunawar, M., and K. Low. 2014. Corporate social responsibility and sustainable development: Trends in Asia. In ed. K. C. P. Low, S. O. Idowu and S. L. Ang, 173–196. Cham: Springer International Publishing.

Asia Monitor Resource Center. 2012. *The reality of corporate social responsibility: Case studies on the impact of CSR on workers in China, South Korea, India and Indonesia*: Asia Monitor Resource Centre.

Bauer, A., and M. Thant (eds.). 2010. *Poverty and sustainable development in Asia: Impacts and responses to the global economic crisis*. Mandaluyong City: Asian Development Bank.

Carew-Reid, J. (ed.). 2009. *Strategies for sustainability: Asia*. London: Earthscan.

Chaponnière, J.R. 2012. The socioeconomic consequences of the global crisis on Asia. In *Socioeconomic outcomes of the global financial crisis: Theoretical discussion and empirical case studies*, ed. U. Schuerkens, 191–221. Oxon: Routledge.

Chapple, W., and J. Moon. 2005. Corporate social responsibility (CSR) in Asia: A seven-country study of CSR. *Business and Society* 44(4): 415–441.

Chapple, W., C. Herzig, and R.C. Slager. 2014. The dynamics of corporate social responsibility in Asia: A 6 country study. *Academy of Management Proceedings* 2014(1): 16813.

Corsetti, G., P. Pesenti, and N. Roubini. 1998. *What caused the Asian curency and financial crisis? Part I: A macroeconomic overview*. Cambridge, MA: National Bureau of Economic Research.

Fukikawa, K., and J. Moon 2004. A Japanese model of corporate social responsibility? A study of website reporting. *Journal of Corporate Citizenship*, 16(October), 45–59.

Fukushi, K., K.M. Hassan, R. Honda, and A. Sumi (eds.). 2010. *Sustainability in food and water: An Asian perspective*. London: Springer.

Gill, S.S., L. Singh, and R. Marwah (eds.). 2010. *Economic and environmental sustainability of the Asian region*. New Delhi: Routledge.

Kotler, P., H. Karatajaya, H.D. Huan, and S. Liu. 2008. *Rethinking marketing: Sustainabe marketing enterprise in Asia*, 2nd ed. Singapore: Prentice Hall.

Laruelle, M., and S. Peyrouse. 2012. The challenges of human security and development in central Asia. In *The security-development nexus: Peace, conflict and development*, ed. R. Amer, A. Swain, and J. Öjendal, 137–160. London: Anthem Press.

Lin-Heng, L., G. Ofori, M.-L.L. Choo, C.R. Savage, and T. Yen-Peng (eds.). 2010. *Sustainability matters: Environmental management in Asia*. Singapore: World Scientific Publishing.

Lipovsky, I. 1995. The deterioration of the ecological situation in central Asia: Causes and possible consequences. *Europe-Asia Studies* 47(7): 1109–1123.

Low, K.C.P., S. Idowu, and S.L. Ang (eds.). 2013. *Corporate social responsibility in Asia*. London: Springer.

Matten, D., and J. Moon. 2008. 'Implicit' and 'explicit' CSR: A conceptual framework for a comparative understanding of corporate social responsibility. *Academy of Management Review* 33(2): 404–424.

Mishkin, F.S. 1999. Lessons from the Asian crisis. *Journal of International Money and Finance* 18(4): 709–723.

Moon, J., and X. Shen. 2010. CSR in China research: Salience, focus and nature. *Journal of Business Ethics* 94(4): 613–629.

OECD. 2001. *Towards Asia's sustainable development: The role of social protection*. Paris: OECD Publication Service.

OECD. 2012. *OECD environmental outlook to 2050: The consequences of inaction*. Paris: OECD Publishing.

Ong, M. 2008. *Contextualising corporate social responsibility in Singapore*. Singapore: Centre on Asia and Globalisation, Lee Kuan Yew School of Public Policy, National University of Singapore.

Petri, P., and T. Vinod. 2013. *Development imperatives for the Asian century*. Development Bank Economics Working Paper Series No. 360. Retrieved from http://ssrn.com/abstract=2295823.

Redding, G., and M. Witt. 2009. China's business system and its future trajectory. *Asia Pacific Journal of Management* 26(3): 381–399.

Rock, M.T., and H. Bonnett. 2004. The comparative politics of corruption: Accounting for the East Asian paradox in empirical studies of corruption, growth and investment. *World Development* 32(6): 999–1017.

Rock, M.T., D.P. Angel, and T. Feridhanusetyawan. 1999. Industrial ecology and clean development in East Asia. *Journal of Industrial Ecology* 3(4): 29–42.

Rowley, C., J. Saha, and D. Ang. 2012. *Succeed or sink: Business sustainability under globalisation*. Oxford: Chandos Publishing.

Smith, D.V., and K.F. Jalal. 2000. *Sustainable development in Asia*. Manila: Asian Development Bank.

The World Bank. 2012. *Turn down the heat: Why a 4 °C warmer world must be avoided*. Washington: The World Bank and the Potsdam Institute for Climate Impact Research and Climate Analytics.

Tipton, F. 2009. Modelling national identities and cultural change: The Western European, Japanese and United States Experiences compared. *International Journal of Cross Cultural Management* 9(2): 145–168.

WBCSD. 2013. *Vision 2050: The new agenda for business: World business council for sustainable development*.

Welford, R. 2005. Corporate social responsibility in Europe, North America and Asia. *Journal of Corporate Citizenship* 17(1): 33–52.

Whitley, R. 1992. *Business systems in East Asia -firms, markets and societies*. London: Sage.

Williams, G. (ed.). 2011. *Responsible management in Asia: Perspectives on CSR*. New York: Palgrave Macmillan.

Williams, C., and R. Aguilera. 2008. Corporate social responsibility in comparative perspective. In *Oxford handbook of corporate social responsibility*, ed. A. Crane, A. McWilliams, D. Matten, J. Moon, and D. Siegel, 452–472. Oxford: Oxford University Press.

Wong, L. 2009. Corporate social responsibility in China: Between the market and the search for sustainable growth development. *Asian Business and Management* 8(2): 129–148.

WWF. 2012. *Living planet report 2012: Biodiversity, biocapacity and better choices*. Gland/London: Worldwide Fund For Nature.

Zhao, L. 2014. Convergence of East-west business management philosophy: The significant development of Chinese CSR theory and practices. In *Corproate social responsibility and sustainbility: Emerging trends in developing economies*, vol. 8, ed. G. Eweje, 79–116. Bingley: Emerald.

Chapter 2
Catalyzing Social Innovation – How Intel Helped to Create a Vibrant Social Ecosystem in China

Barbara Igel, C.Y. Yeung, and Sheikh A. Prince

Origins of Intel's Approach to CSR in China

In the early 1990s, the Chinese information and communications technology (ICT) industry was still in its infancy. It was by no means certain that China would become the global leader in ICT production, consumption, and R&D that it is today, despite the government's strategic intent to promote the local industry. At the time, the global ICT industry was dominated by vertically integrated companies, confining the nascent PC industry to a handful of large multinational corporations in China.

Intel recognized that to grow its business in China, it had to first help China to grow. Under the stewardship of the then-CEO and co-founder Andy Grove, Intel's China strategy focused on enabling a vibrant horizontal computing industry based on open standards. Intel began systemically engaging local stakeholders across the industry, leveraging its experience and resources to help local companies "crawl, walk, and run". As part of this strategy, Intel first established its fully owned operation and legal entity, the Intel Architecture Development Lab, in 1994 in Shanghai with the sole purpose of enabling a vibrant computing ecosystem in China. Intel brought in teams of technologists and business experts from around the globe, developed local talent and worked with local developers to create advanced hardware and software products based on Intel's architecture. It established a software group to help the software development community to master the best-in-class consumer and business applications based on the latest technology. This would ensure that Chinese software applications were available at the launch of Intel's latest

B. Igel (✉) • S.A. Prince
Asian Institute of Technology, Pathumtani, Thailand
e-mail: igel@ait.asia

C.Y. Yeung
Intel China Ltd., Beijing, China
e-mail: c.y.yeung@intel.com

G. Lenssen et al. (eds.), *The Role of Corporate Sustainability in Asian Development*, Advances in Business Ethics Research 7,
DOI 10.1007/978-3-319-45160-2_2

Fig. 2.1 Intel's business growth in China since 1985

generation of PC by its OEM customers (this was the precursor of Intel's 8000 strong Intel Software and Services Group today). Intel also organized multiple large-scale matchmaking events to bring global and local hardware and software players together and form partnerships to jump-start the Chinese computing industry development.

Being Intel's new home country, Intel was abiding by rules that were stricter than China's local regulations and laws. Intel also had closely followed the government agenda since setting up its operations in China in 1985. When the government agenda targeted the west part of China, Intel invested in a major factory at Chengdu (see Fig. 2.1). A few years ago when the government talked about how to revitalize the northern and eastern parts of China – Intel set up its first fabrication factory in Dalian. In line with the government agenda to upgrade the industry, Intel has also invested into a regional R&D company in Shanghai.

Intel also brought its flagship Intel Development Forum to China to introduce the latest and best-in-class technology to local developers. And in 1998, Intel Capital which is Intel's strategic venture capital arm began operating in China to further stimulate and accelerate the computing and communication industry's growth.

These collaborations with companies, policymakers, academics, and end users across the industry, provided Intel real-time knowledge and insight into the local computing industry and its markets and enabled ideas to quickly cross-pollinate, risks to be shared, and innovation to thrive. More importantly, these partnerships played a key role in stimulating an environment for collaborative innovation based on open industry standards, which catalyzed the development of a vibrant ICT sector that laid the foundation for Intel's success in China today.

By the turn of the century, China's economy had grown so big, that many multinational corporations (MNCs) designated it as an independent geographic segment. MNCs typically tried to penetrate a large foreign market by exploiting their connections with local key stakeholders. But Intel, instead of regarding China simply as a marketplace, aimed to be a part of China's economic growth and development by trying to understand what China's dream was about – achieving national rejuvenation, prosperity and wellbeing of people. The third plan of Chinese national conference focused on the social and political issues of the country towards revivalism. Intel had anticipated this quite some time ago and started to review and reformulate its CSR strategy to create value in support of China's Great Dream.

By 2012, according to the International Data Corporation (IDC), the PC industry in China had surpassed that in the U.S. to become the world's largest market mostly with Intel Inside. As Intel's strategy evolved, it continued to promote the smart use of ICT in China as a tool to drive China's economic transformation and social development. Not only had the ICT industry dramatically grown itself, it had enabled tremendous growth in other industries. Intel recognized that its contribution was not just in the way that technology enabled innovative solutions, but in how enabling the development of the technology ecosystem stimulated the entire industry, fostering and multiplying the impact of those solutions. Similarly, new ways of thinking – social innovation- just like new types of technology, could also be applied to enable new social solutions. Inspired by this, Intel began to explore how the very same approach of building an enabling ecosystem and fostering cross-sector collaboration might be applied to addressing China's prevailing social and environmental challenges. But to achieve this, it was crucial to build a vibrant "social ecosystem" to develop, test, use, scale, and share new approaches to solving social problems.

As the Chinese government began to shift its direction toward more inclusive and sustainable development, and with Intel's experience in China, it was clear to Intel that to solve social problems China would need to build up its social infrastructure as aggressively as it had built its technology infrastructure to grow the ITC industry. And by following a similar approach, Intel could help China create a vibrant social ecosystem to stimulate social innovations that could tackle China's rising social and environmental challenges. From technology innovation to social innovation, catalyzing collaborative solutions by creating ecosystems around key social issues could be the recipe for China to fulfill its "China Dream". But this required businesses to play their part by redefining CSR and embracing a new model focusing on real sustainability.

On 29th November, 2013, Mr. CY Yeung, the Director, CSR, Intel China presenting his case at the ABIS Global CSR Conference at the Korea University, in Seoul explained to the audience:

I was assigned to lead a particular role across organization. I was asking myself questions that were coming from the business and also from the technology world, such as 1) what impact are we at Intel trying to make? 2) Who are our key stakeholders? 3) What values does CSR really target? CSR is not something that is already carved in stone, we rather find it in our own job description.

The Evolution of CSR in China

CSR has been evolving rapidly in China over the last 5 years, in part from a more discerning middle class of consumers, and in part from increased pressure from traditional and social media. Historically, Chinese businesses regarded CSR as philanthropy. But, as the government realized that China could no longer afford to grow at the expense of social stability and environmental sustainability, Chinese businesses also began to identify CSR as their contribution to addressing these more complex issues. Now it is clear the current Chinese leadership has elevated CSR to a national priority by driving smart, inclusive, and sustainable development for a "Beautiful China" (as China's new leadership declared in 2013).

Despite the heightened attention, social challenges such as education, health care, elderly care, pollution, and the wealth divide are so complex that no single government or organization can tackle them on its own. Cross-sector and cross-border collaboration is required among policymakers, businesses, nonprofits, academia, development agencies, and citizens around the world. In China, there is a unique opportunity to leverage the support of the Chinese government to create an enabling environment for collaboration and innovation to flourish.

Over the last 30 years, multinational companies have played an indispensable role in advancing China's economic development through foreign direct investment, technology transfer, local talent development, and created shared economic value both for China and for corporate shareholders. With the business landscape now rapidly shifting away from an investment-driven, resource-intensive, export-led growth model, the real opportunity for multinational companies is to move up the value chain to create shared economic and social value to grow together with China. As such, it is imperative for corporate social responsibility to embrace the opportunity of corporate social innovation.

Evolutionary Learning: Intel's Journey to CSR 3.0

The development of Intel's CSR strategy in China has been typical of many multinational companies. Nevertheless, it has been a long but exciting journey. This thinking initially started at the community level where Intel strove to be an asset to the communities in which it was operating. This initial program laid a solid foundation for a range of corporate volunteerism, philanthropy, and social contribution programs (CSR 1.0).

In 2000, Intel identified the need to improve the quality of education and launched the *Intel Teach* program in China with the Ministry of Education. Intel recognized that it could leverage its core competencies in innovation and technology to help China develop new ways of effectively delivering quality. Intel believed that education is key to China's economic and social development, as well as to cultivating the next generation of innovators. Improvements in education create the educated workforce necessary for a competitive ICT industry, and boosts innovation. This, in turn, stimulated demand for computing and communication devices, benefitting Intel's business.

Research has shown that the degree of ICT usage correlates directly with economic and social development indicators, and therefore the broader use of ICT would enhance national productivity and economic development (Kraemer and Dedrick 2001; Qingxuan and Mingzhi 2002). In short, by focusing on education, Intel would not only help contribute to China's education goals and its economic and social development, but also would grow China's demand for ICT products. This approach, often called "shared value" nowadays, is considered by Intel as CSR 2.0. The concept of shared value can be defined as policies and operating practices that enhance the competitiveness of a company while simultaneously advancing the economic and social conditions in the communities in which it operates (Porter and Kramer 2011).

Intel's CSR team recognized it would require a paradigm shift on the part of businesses to embrace corporate social innovation. Intel's CSR agenda identified the need to drive social innovation and thus emphasized building the ecosystem that catalyzes social innovation as the path to sustainable social development and long-term business value creation. CSR teams would require a mandate to create shared business and social values to achieve a systemic impact. The goal would be to create a new driving force for China's development, one that would be more sustainable than in the past. Intel believed this concept of taking an issue-centered approach and building an ecosystem of organizations to address the issue represented a new way of thinking about CSR and coined it CSR 3.0.

Intel's CSR 3.0 strategy calls for working with other stakeholders from all sectors to better identify the root causes of social issues, identify effective partnerships, deploy resources more effectively at a cross-sector level, and strive for systematic solutions for collective impact. CSR 3.0 initiatives are always interesting among the business academia, policy makers, governments, NGOs- not only locally but also globally- as it provides new opportunities for expanding the market and creating future product solution.

Definition – Intel's CSR 3.0

Intel's CSR 3.0, is defined by a fundamentally different approach to value creation which has important implications for business and society in the twenty-first century. This model of collaboration can mobilize resources at sufficient scale to deliver long-term social change by expanding the scope of the conventional corporate responsibility definition from "corporate-focus" to "regional-focus" by creating an ecosystem of stakeholders where ideas can be cross-pollinated and resources can be pooled to tackle social problems together. Creating innovative solutions to problems in society unleashes new market demands, but achieving these CSR dividends requires creating an environment that excites key influencers, revitalizes the organization, and ignites the marketplace. Intel's CSR 3.0 emphasizes building the ecosystem that catalyzes social innovation as the path to sustainable social development and long-term value creation.

Catalyzing Social Innovation Through CSR 3.0

Corporate innovation tends to be associated with the creation of new products or processes that deliver value to the customers in new and improved ways. *Social innovations then are new ideas (products, services and models) that simultaneously meet social needs (more effectively than alternatives) and create new social relationships or collaborations (Murray, Calulier-Grice and Mulgan,* 2010*).* Social innovation is the process of inventing, securing support for, and implementing novel solutions to social needs and problems. These solutions are both social in their ends and in their means. As with technological innovation, social innovation is thus both an approach and a result: a way of thinking and the application of that thinking to solving social problems.

As the Chinese government began to shift its direction toward more inclusive and sustainable development, it was clear to Intel that to solve social problems China would need to aggressively build out its social infrastructure, just as it had built out its technology infrastructure to grow the ICT industry. By following a similar approach, Intel could help China create a vibrant social ecosystem to unleash social innovation to tackle China's social and environmental challenges.

The CSR 3.0's vision is to catalyze that process, not only as a proactive actor involved in driving change directly, but more importantly, by contributing to the creation of an ecosystem that is conducive to the formation and realization of those ideas. Through active participation in the development and implementation of that social innovation, the company not only achieves its economic goals, but also ensures its own long-term sustainability by becoming an indispensable component of society.

The Evolution of CSR

The CSR of today rethinks the concept of value creation and the role of collaboration, mobilizing resources at sufficient scale to deliver long-term social change. Originally, leading CSR practices asked, "How can our company best make a difference using our unique competencies?" The next stage asked, "How can our company strategically partner with others to address the problem?" Now a more systemic approach has evolved requiring a better understanding of the problem, the overarching solution, and the role of different stakeholders as a part of that solution. This now requires companies to ask, "How can we inspire and drive progress from all key stakeholders that could currently or potentially influence the issue in the future?" Crucially, focusing on the issue first changes the perspective and reveals new options for how a company can add value as one of many contributors to the solution: for example, through utilizing corporate networks of suppliers and vendors, its relationships with government, its employee skill sets, its problem-solving abilities, and so on. Creating an ecosystem and "learning network" among all the relevant stakeholders addressing the issue allows ideas to quickly cross-pollinate and resources to be pooled to tackle social problems together.

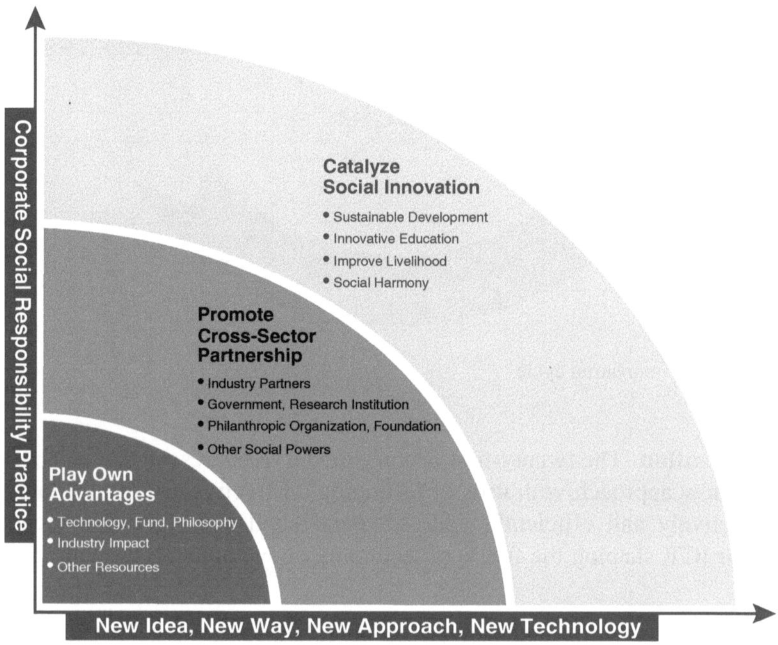

Embracing the "WE" vs "I" paradigm of CSR helps create a bigger pie for business and society to collaborate and thrive for collective impact. This helps stakeholders to gain better insight into the root-causes of social issues as opposed to the symptoms, and to more effectively and efficiently deploy resources at a cross-sector level than at a single issue level which is critical for systematic change.

The innovation cycle (see Fig. 2.2) defines Intel's relationship between its business, its industry, and society. It is a core concept of CSR 3.0. Driving innovation in any one of these three domains impacts the others in mutually supportive ways.

Technology Innovation is the Raison d'être for Intel and what Intel stands for. Intel's founders helped spawn the ICT industry through transistor circuit. Moore's Law, proposed by Intel co-founder Gordon Moore, has contributed to productivity gain by delivering more function at the same cost – not only in the ICT industry but also the entire economy.

Collaborative Industry Innovation enables a vibrant open industry ecosystem based on open standards, lowers the industry entry barriers and encourages broad scale innovation. For Intel, technology innovation and industry innovation come hand in hand.

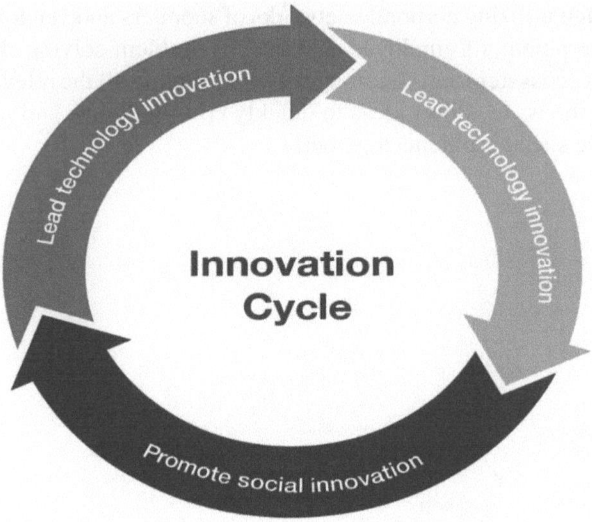

Fig. 2.2 INTEL's innovation cycle

Social Innovation The twenty-first century model of addressing social challenges requires a new approach with uses of technology. Effective use of technology can more effectively and efficiently address social challenges and also create new markets for ICT, shaping the future of technology development, propelling a virtuous circle.

Intel started by identifying unsolved social problem as a way to orient its technology, product and market strategy. The resulting innovation, in turn, provides new value within industries, which then can create social value and social innovation as end-users apply the innovations in ways that address social issues. As social solutions improve and alter the needs coming from society, society then requires a new and refreshed orientation for technology innovation, reinvigorating the innovation cycle anew. Intel sees the innovation cycle as an effective strategic framework that acts as a compass to orient the logic and goals of cross-sector partnerships (Fig. 2.3).

CSR 1.0 is primarily focused on reducing negative impacts to society and supporting social issues through philanthropic or social contribution programs at a community level.

CSR 2.0 takes a more strategic and **collaborative** approach; companies work with value chain members to identify win-win opportunities where their core competencies can **create shared value** for society and themselves, or that are closely related to their business.

CSR 3.0 takes into account a company's **broader ecosystem**, and works to strengthen that ecosystem to provide systemic solutions to social problems through **cross-sector partnerships** for long-term and **fundamental improvements to society**.

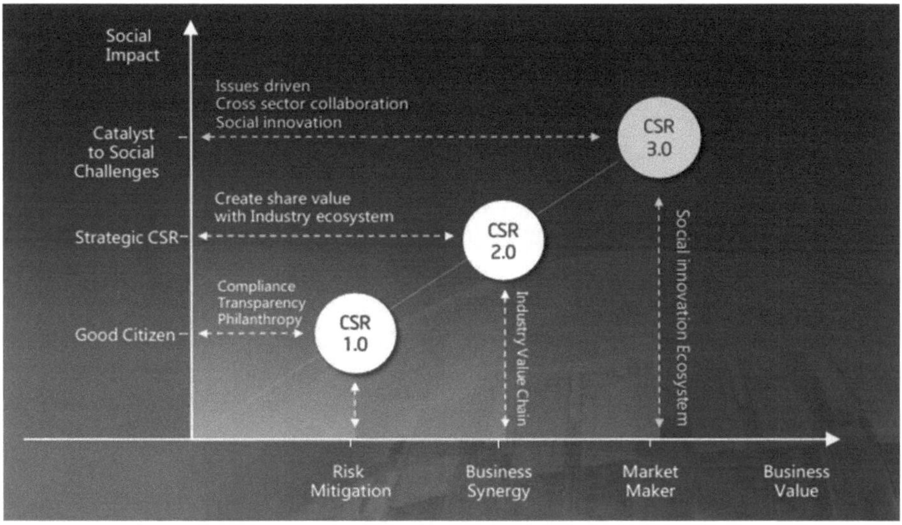

Fig. 2.3 Strategic CSR creating social impact & business value

The Intel group is all about technology innovation and depends on its customers' success and they are facing more and more challenges concerning the environment, education, supply chains, talent acquisition and so on. The government's agenda on public health care, urbanization, modern agriculture etc. shown on the left side of Fig. 2.4 opens up numerous possibilities for contributions from corporate CSR initiatives.

The social ecosystem platform shown in the middle, connects between society's present needs and anticipated future needs while the core engines powering the business, the technology development and economics are shown on the right side. In this area are numerous opportunities for social innovation, where academicians, policy makers, government, NGOs, business leaders and innovators can work together to make innovations happen that create shared value.

Intel created value for both, its business and for its shareholders by mobilizing different sectors across the industry and connecting best practices, experts, business leaders and politicians to work together on the issues of high importance to the society. That requires leadership, collaboration and innovation (Fig. 2.5).

In late 2010, Yin Gefei, Vice President of WTO Tribune approached Intel. He was intrigued by Intel's approach to CSR and its thinking about the CSR ecosystem. He sought to further explore and elaborate on the concept, and share this way of thinking with other businesses in China. Yin Gefei and Yu Zhihong, Chief Editor of WTO Tribune, had long been strong advocates of the CSR movement in China. Intel was sharing the same aspirations to contribute to China's inclusive and sustainable development through its own experience. This marked the beginning of the journey to publish the Intel China case story. Two years later, on Christmas Eve in 2012, in a coffee shop near the Olympic Park, C Y Yeung, Director, Corporate Responsibility,

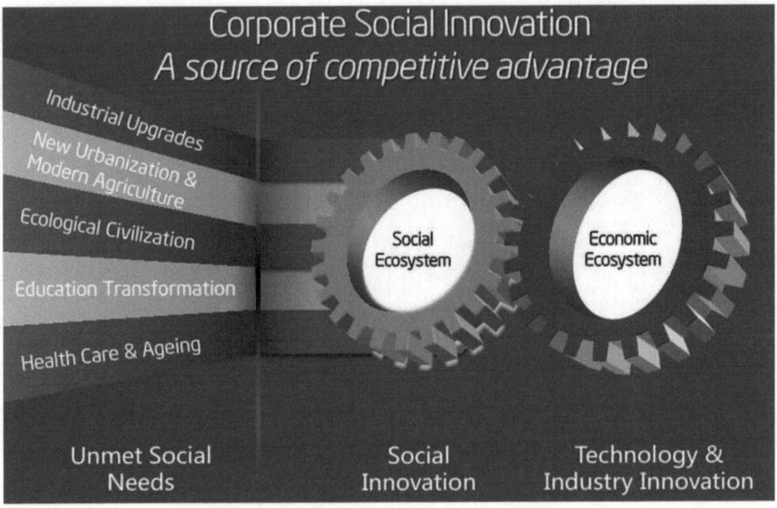

Fig. 2.4 Corporate social innovation creating competitive advantage

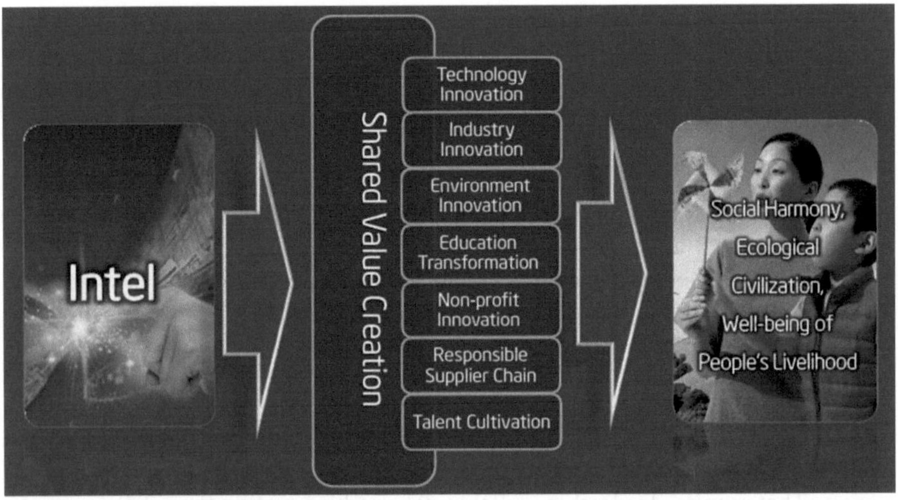

Fig. 2.5 Shared value creation

Intel China analyzed previous experiences and mapped out an outline. Much of this thinking around the social ecosystem had grown out of Intel's earlier work on enabling technology and the ICT industry ecosystem starting from the turn of the millennium and Intel's pioneering work since 2009 around social innovation.

Examples of Intel's social innovation initiatives, such as the Intel Teach program, ICT in medical services and caring for the elderly are described below.

The Intel Teach Program

The Intel Teach program launched in China at the beginning of 2000 provides a good example for shared value creation. The aim was to accelerate China's education quality to areas and regions most in need. Intel believed that education is key to China's economic and social development, as well as to cultivating the next generation of innovators. Intel began to focus on education as an area that would generate multiple social "wins" and boost innovation in the economy that would, in turn, create demand for computing and communications. Broader use of ICT would enhance national-wide productivity and economic development. The high-tech ICT industry would also need an well-educated workforce if it was to experience rapid development that Intel envisioned. Increased computing and communication usage correlates with economic and social development. In short, by focusing on education, Intel would not only help contribute to achieving China's domestic goals, but also help create new uses and uses for ICT, creating future demand for Intel's products.

The Intel Teach program has trained over 2.2 M teachers in the twenty-first century. Through this collaboration, the government also invested an additional RMB 20 billion to establish over 1000 schools and 100 colleges specifically located in impoverished areas in China. The program's e-learning model that evolved from the collaboration has changed thinking about the potential of ICT in the China's education system. It is transforming the education experience in many provinces.

Intel's collaboration on China's rural education is more than just a CSR program. It is a coordinated and long-term attack on China's rural education deficiencies in partnership with key influencers. This approach is especially important during the rise of China that, while lifting hundreds of millions of Chinese out of poverty, is also producing great strains on natural and human systems.

Intel's role in this multi-sector partnership has won it the prestigious "Outstanding Contribution to Education" award 8 years in a row by the Chinese Ministry of Education. The effort has also created new market demands. Because of the notable success of the Intel Teach program to utilize ICT as a training tool, as well as the new teaching methods and models that it provided, Chinese schools are now increasingly equipping themselves with computers and digital tools – making a direct market impact. And in other cases, education stakeholders are seeing the effectiveness of ICT as a teaching tool and coming up with even more and better uses for ICT in the classroom.

Furthermore, the students benefiting from the program and/or application of technology and now are more likely to have a greater earning potential from their improved education, which may translate into more demand for ICT products that they otherwise would not have been able to buy. It summary, the Intel Teach program has become a positive reinforcing cycle that applies new teaching tools and techniques, but also leads to even better education innovations and ways to use more ICT in the classroom. In other words, this coordinated effort to address education deficiencies in China is unleashing wave after wave of innovation to address education challenges through new technologies that is creating market demand for the ICT industry. Looking to make an even bigger impact on education in China, Intel began to shift gears towards driving educational transformation at a system level through policy advocacy, professional development and teacher-training with a broad range of cross-sector partnerships – creating an ecosystem where stakeholders can collaborate to create meaningful and lasting change. This paradigm shift lifted Intel's engagement on education from working within limited partnerships to catalyzing lasting structural change through policy advocacy and deeper and broader collaborations.

Over time, Intel began to shift the Intel Teach program in China toward driving education transformation at a systemic level through advocacy with policymakers, professional development, and teacher-training, and a broad range of cross-sector partnerships—creating an ecosystem for stakeholders to collaborate on and create meaningful and lasting change nationwide. This paradigm shift built on the lessons learned from Intel's work with the ICT industry a decade earlier and elevated Intel's work on education from teacher training alone to catalyzing lasting structural change. It ensured Intel was an indispensable partner in the industry, providing Intel with a unique long-term competitive advantage. Intel has been applying the same approach to collaboration in China across a range of other issues, such as medical care, aging populations, and the low carbon economy.

Innovation Initiative for Nonprofits

When a devastating earthquake hit Wenchuan in Sichuan province on May 12, 2008, for the first time China's nascent nonprofit sector played an important role in providing disaster relief, bringing complementary resources to those provided by the government. It became clear to Intel that nonprofits had the potential to do much more, and not just in disaster relief. In addressing its many social and environmental issues, China needed to develop a vibrant social sector to create systemic impact. In early 2010, under the auspices of the Charity and Welfare Department of the Ministry of Civil Affairs (MoCA), Intel, together with the China Foundation for Poverty Alleviation (CFPA) and the Narada Foundation, launched a major initiative, the Innovation Initiative for Nonprofits (IINP).

The goal of IINP was to identify, support, promote, and create a community of leading social innovators. Learning from Intel's previous experience with stimulating the ICT industry, IINP became a focal point in bringing together stakeholders to collaborate, helping individual organizations grow and raising the profile of the whole sector. As of today, the initiative has connected over 900 leading social organizations, academics, and businesses, and has evolved to become a learning network promoting best practices, a platform for collaboration among nonprofits and social enterprises, and an opportunity for academics to undertake research and for policymakers to gain insight into and test out policy ideas. In short, it has evolved into an open social innovation lab, a "think and do tank" for collective impact, giving inspiration to a new "WE-centric" ecosystem model of CSR, i.e. CSR 3.0.

Helping China's Medical Sector

In the medical sector, Intel promotes the ability to share medical information in an accurate and timely manner amongst medical practitioners in China. By partnering with the Ministry of Health, Intel helped to build and improve the medical information systems, including establishing a digital medical-record standard. Intel partnered with 30 medical application service providers, and has supported over 330,000 village health-care centers to develop their medical information capacity throughout China utilizing Intel's technology and that helps them access crucial medical information more timely and accurately.

Contribution to Low Carbon Management

To contribute to China's 2020 carbon intensity goal, Intel established the *Digital Energy Solution Center* (DESC) in partnership with other companies and industry associations in China. The DESC brings together the private sector, non-profits, government, consumers, and other stakeholders to find ICT solutions to energy conservation, low carbon development, and sustainable growth. DESC has publishes research reports to identify the challenges and potential contribution to low-carbon development and growth in China, such as the 2011 report titled "Information Communication Technology Promotes China's Low Carbon Economic Development".

Improving Care for the Elderly

In 2012, Intel China published a report on "Sustaining the Golden Thread: An Economic Approach to Aging for China". Intel also co-hosted the Aging Care Service Seminar to discuss among h key stakeholders, how best-practice ICT solutions can be a tool to meeting the needs of elderly, such as preventing chronic diseases and improving relationships with families and friends. The initiative aims to facilitate cooperation among China's policy makers, researchers, and business leaders, and to trigger more in-depth thinking about the potential role of ICT to address aging-related problems and potential opportunities. Internally, Intel also established a dedicated business organization to look into such issues and explore opportunity to leverage ICT to help the elderly live better lives.

Social innovation was recently being driven by Cinnovate, an independent NGO incubated by Intel after extensively engaging with other leading CSR experts in China to get their feedback on this concept. In particular, Chen Feng, Deputy Director, Research Bureau of State-owned Assets Supervision and Administration Commission, Guo Xiuming, Director of the Legal and Policy Division, Ministry of Industry and Information Technology (MIIT), Professor Yang Dongyu, Beijing Guanghua School of Management, Li Weiyang, Director of CSR Division, State Grid, and Wen Li, CSR Expert from PetroChina (CNPC) provided invaluable feedback that has been shaping Intel's evolving thinking and guiding the future path of corporate CSR in China.

The Value of CSR 3.0 to Intel

The focus on corporate responsibility creates value for Intel and its stakeholders as it helps Intel manage its business more effectively and identify ways to apply our technology and expertise to benefit the environment and society, which in turn helps it mitigate risks, reduce costs, protect brand value, and identify market opportunities. Over the years, Intel's efforts have commanded a strong following in China. Many case studies have been written and Intel has been invited to various government affiliated committees and association to shape the CSR evolution in China. Intel's Global Reporting Initiative (GRI) based China focused CSR report has been sought after and was used by leading MBA schools. Intel has been playing an active role across China, from Beijing to Shanghai to Chengdu and Dalian. Intel believes that MNCs have a lot to contribute to their host countries beyond, investment, technology, jobs and tax revenue. Given its global presence and touch points across the various sectors, Intel can play a key role in China's smart, inclusive and sustainable development. If a MNC can accomplish these goals, reputation, and the license to operate will be a moot point.

Intel believes that it can apply its technology and experience to help improve energy efficiency, address critical environmental challenges such as climate change, and improve education access and quality worldwide. Designing products with improved energy-efficient performance helps it meet customer needs and identify market expansion opportunities; improving energy efficiency in its operations helps it reduce the emissions and energy costs; and investing in training, diversity, benefits programs, and education enables the company to attract and retain a talented workforce.

Intel's investments in education expand opportunities for young people while also benefiting the company. Education is the foundation of innovation, and as a technology company, Intel's success rests on the availability of skilled workers, a healthy technology ecosystem, and knowledgeable customers. In turn, the health of local economies—including those where Intel's employees live and work—depends on access to technology and quality education. Intel's education programs support its long-term corporate diversity objectives by encouraging girls, women, and students in underserved communities to pursue careers in science, technology, engineering, and math. Applications of technology in education also create market opportunities for Intel.

CSR 3.0: Implications for Practitioners in China Today

CSR has been evolving rapidly in China over the last 5 years, in part from more discerning middle class consumers, and also in part from increased media pressure, as shown during the 2008 Wenquan earthquake when the media put corporate donations directly under the spotlight. Throughout the early 2000's, Chinese corporations tended to interpret CSR mainly as philanthropy or social contribution programs. But the government realized more and more that China could not continue to afford to grow at the expense of social stability and environment sustainability. The 12th 5-year plan finally mandated balanced and "harmonious" development, and the current leadership now elevates social responsibility to a national policy level by dictating a smart, inclusive and sustainable development model for a "Beautiful China". This new attention to the social impacts of business is leading to a deeper understanding of CSR for stakeholders in China, as can be seen in areas such as new government standards to dramatic increases in CSR reporting by the private sector.

Despite the recent increased attention to CSR, social challenges such as such as education, health care, elderly care, pollution, and the wealth divide are so complex that no single government or organization are able to tackle it on its own. Cross sector and cross boarder collaboration is required among policymakers, businesses, non-profits and academia, and development agencies locally and globally. In China, there is a unique opportunity to leverage the support strong and progressive governments to create an environment for collaboration and for innovation to flourish.

Over the last 30 years, MNCs have played an indispensable role in advancing China's economic development though direct investment, technology, cultivating local talent, and creating shared economic value both for China and for corporate shareholders. With the business landscape now rapidly shifting away from a resource intensive export-led growth model, the real opportunity is to move up the value chain to create shared economic and social value to unleash another 30 years of development, providing corporations a new opportunity to grow together with China. This is what CSR should be all about.

But as a company with operations in China, what are the implications for business leaders and practitioners in China? These three key questions below can be helpful to start the internal conversation about how to harvest the "CSR Dividends" within the unique sociopolitical context of China:

- How to align corporate vision and mission with prevailing government imperatives, taking economic, social and environmental factors into considerations, and successfully revitalize the organization, excite key stakeholders and ignite the marketplace?

- How to identify and explore opportunities to create sustainable shared business and social value to support market access, market development, license to operate, reputation, and to incubate a vibrant internal culture?
- How to revamp the corporate structure (hardware) and processes (software) to create an environment for collaboration and innovation to flourish?

Five Steps to Launch a CSR 3.0 Strategy

For most corporations new to CSR, or aiming to rethink how concepts of sustainability and corporate responsibility can boost their competitive advantage, the journey of internal cultural-change and strategy reorientation can be an exciting one, yet it also can be daunting.

Based on Intel's experience the 5-step framework shown below may help to maximize the speed and effectiveness of a CSR change-program by providing a conceptual roadmap aligned with the principles of CSR 3.0.

1. Establish a CSR champion or group that is part of the organization's senior management and strategic management. Empower the champion as a change agent to explore and test innovative approaches that can transform the business model in line with societal expectations and opportunities.
2. Identify the most pertinent societal issues to focus on in your host/home county and analyze where these align with the strategic intent and core capabilities of the organization that could maximize shared business and social impact. Two key considerations that should be discussed internally are:
   - How to align the corporate vision and mission with prevailing societal and governmental imperatives, taking into account economic, social and environmental factors.
   - How to identify and explore opportunities to create sustainable shared business and social value to support market access, market development, license to operate, and enhance reputation.
3. Map the issue-specific stakeholder ecosystem and convene stakeholders across social organizations, academia, government, business, media, and others.
4. Dismantle boundaries between sectors by mobilizing and rallying both internal and external stakeholders in the ecosystem to collaborate together and innovate to develop effective solutions.
5. Create an open feedback platform to capture real-time learning from rapid experimentation and scaling in the field to guide further efforts.

Concluding Thoughts

Across the globe, strains on social and environmental systems are causing increased volatility across borders and impacting quality of life. It is time to rethink how we collaborate and innovate. It is time for a paradigm that looks far enough into the future and that is sufficiently broad to cross sectors and to realize lasting structural change. The CSR 3.0 paradigm holds the promise that it can achieve required change at scale. It is in the best interest of all to consider the lessons and advantages of re-imagining how corporations relate to society, and how corporate social innovation can be a strategy and path to business success and sustainability.

This case chapter attempted to describe Intel's journey of corporate social innovation in China as well as the latest thinking and practices. It is intended to inspire more discussion, debate, and engagement with like-minded fellow travelers. The societal, environmental, and economic issues are so serious, and the business opportunities for creating solutions so immense, that stakes are now higher as ever.

Acknowledgement This case was written with the full support of Intel's top management who provided us access to all relevant information about Intel's China experience. Special thanks go out to C Y Yeung, Director, Corporate Responsibility, Intel China without whose encouragement and dedication this could not have possible.

References

Kraemer, K.L., and J. Dedrick. 2001. *Information technology and productivity: Results and policy implications of cross-country studies*, 257–279. Oxford: Oxford University Press.

Meng, Qingxuan, and L. Mingzhi. 2002. New economy and ICT development in China. *Information Economics and Policy* 14: 275–295.

Murray, Calulier-Grice, and Mulgan. 2010. *Open book of social innovation, March, cited in European Commission: Social innovation.* http://ec.europa.eu/enterprise/policies/innovation/policy/social-innovation/index_en.htm. Accessed Apr 2015.

Porter, Michael E., and Mark R. Kramer. 2011. Creating shared value, Harvard Business Review, January–February, 2011.

Chapter 3
Lenovo's Venture Philanthropy: Evaluating and Planning

Maria Elena Baltazar Herrera and Steven White

Introduction

With a population of over 1.3 billion people, China is one of the two largest economies globally and also one of the fastest growing. This vibrant economy emerged from a closed, centrally-planned economy when economic reforms were implemented under the leadership of Deng Xiaoping in 1978 (Barne 2012). These economic reforms include opening up the country to foreign investment, transformation and privatization of enterprises, and allowing entrepreneurs to start businesses (Barnc 2012).

The success of China's economic policies resulted in immense changes in the country, but in such a vast country, deep social challenges remain. Particularly serious are those related to water scarcity and quality, industrial pollution, and greenhouse gas emissions, labor conditions, product safety, corruption, increasing rural and urban gap, and inequality and economic exclusion (US Chamber of Commerce – Asia 2012). While these challenges also exist in other countries, they are particularly salient in China.

In this chapter we focus on the distinct situation of China, capturing an important period of time in its transition from a centrally planned economy to greater economic liberalization. This corresponds to the time when China began to engage non-government organizations (NGOs) in development efforts.

M.E.B. Herrera (✉)
Asian Institute of Management, Makati, Philippines
e-mail: mayaherrera@gmail.com

S. White
Department of Innovation, Entrepreneurship and Stratergy, Tsinghua University School of Economics and Management, Beijing, China
e-mail: white@sem.tsinghua.edu.cn

G. Lenssen et al. (eds.), *The Role of Corporate Sustainability in Asian Development*, Advances in Business Ethics Research 7,
DOI 10.1007/978-3-319-45160-2_3

The integration found in China differs between multinational companies and local Chinese companies, which can be attributed to differences in perspectives. It is usual for multinational companies to adhere to global corporate policies and international standards, which in many cases go beyond what is required by local laws. Many Chinese companies, however, are not even able to comply with local regulations. Private firms, in order to enhance their corporate reputation and brand value, have been accelerating the diffusion of CSR practices in China by actively seeking additional social capital from government, communities, and stakeholders (Herrera and Roman 2011).

This chapter presents Lenovo's Venture Philanthropy Program (VPP), Lenovo's efforts to contribute to capacity-building in NGOs in China. Lenovo launched its VPP in 2007 and disbursed its first grants in 2008. Its mission was to provide capability enhancement and financial support to individuals and organizations wanting to start or substantially expand a social enterprise.

The case provides an opportunity not only to evaluate the situation of NGOs in China, but also an opportunity to study the evolution of a Chinese company into a global brand and changes in its attitude towards and implementation of Corporate Social Responsibility (CSR). The chapter focuses on Lenovo as a Chinese corporation with a global brand and how this influenced its effort to expand and develop the space for Chinese NGOs. This case shows how Lenovo's CSR efforts align with China's efforts to strengthen NGOs in order to supplement government development initiatives.

Empirical Study: CSR, China and Lenovo

We undertook a qualitative study of a specific CSR program focused on the underdeveloped NGO space in China and the corporate context of a local firm that has developed into a multinational. The study includes a review of CSR in the PC industry and in China, Lenovo's corporate development, and a description of how Lenovo initiated and managed its Venture Philanthropy Program. It is based on primary data provided by Lenovo, extensive interviews with Lenovo's CSR Manager, and a wide range of academic and media sources.

The following sections present findings related to three main topics. First, an overview of CSR in the personal computer (PC) industry and in China. Second, a description of Lenovo's founding and development to provide an understanding of the firm's internal context. Third, a detailed description of Lenovo's VPP, from its initial design and through subsequent evolution and review.

CSR in the PC Industry

The PC industry is a major driver for growth and innovation worldwide. Manufacturing in the PC industry is knowledge-intensive and requires reliable access to human capital and facilities for research and development and innovation development. Most competitive manufacturing industries are located in countries such as Philippines, Malaysia, Singapore, Taiwan, Korea and China (Schmitz 2006). In the face of growing sustainability issues, companies in the industry are adapting their business models in order to manage risk and create value for their organizations.

The main sustainability issues that are particularly relevant to the PC industry are related to the materials used in production, energy consumption during production, and usage of electronic products. The most important social issues related to this broad sector (information, communication and technology, or ICT) are poor working conditions from mining raw materials, producing and assembling hardware components, and recycling and disposal of electronic waste. The relative importance of issues related to quality of jobs varies according to the region of operation.

For the industry to have a positive impact on such issues, the major companies such as Acer, Apple, Dell, Hewlett-Packard (HP), and Lenovo cannot act in isolation. To be most effective, these companies need to work with their industry competitors to address these social, environmental and industry issues (Standberg 2013). Currently, however, they vary greatly in their interpretation and response to sustainability issues.

Acer, a Taiwanese multinational ICT firm, has launched an inclusive mechanism for sustainable development by analyzing the challenges faced by the industry as well as the commercial implications of these issues for its business. Its core values (speed and efficiency, cooperation and teamwork, and openness and communication) complement its sustainability goal of being a successful global information technology company with a triple-bottom line (Acer Website 2014). Its corporate responsibility policy states:

- The company strives to meet the expectations of its stakeholders.
- Through a top-down approach, the highest leadership takes charge of promoting CSR, as well as mapping out feasible action plans for marketing its products and services.
- Effectively monitors and manages the risks derived from sustainability-related issues through its regional and branch offices, thus making use of inherent opportunities.
- Works side by side with suppliers to promote business ethics, minimize climate risk, and improve resource efficiency.

Acer Foundation was formed to assist the company in pursuing its sustainability effort through targeted programs, including Digital Innovation Awards, Acer CSR Forum, Acer Incredible Green Contest, and long-term thesis awards that nurture young researcher. All Acer Taiwan employees get 2 days of paid volunteer leave per year, allowing them to join the charity groups and community activities that they support and thereby helping their communities' social and environmental needs. Volunteer service allows employees to gain new experiences and ways of thinking.

US-based Apple, in contrast, despite being the world's most valuable brand in 2013 and ranked 6 in 2013 Fortune 500, does not seem to have a coherent strategy when it comes to sustainability and CSR. Prof. Gregory Unruh of Harvard identifies the main issues that Apple has been struggling with in the areas of sustainability and CSR (Unruch 2012).

- The company lacks a stakeholder engagement strategy. The company is reluctant to engage with any stakeholder that has a critical point of view of the business.
- Apple has not adopted the triple bottom line; it has historically focused on maximizing its profits.
- The company does not have a dedicated CSR leadership. Hence there is no team dedicated to driving the sustainability agenda of the company.
- Apple's transparency is still far from satisfactory. The company has not released a sustainability report.
- Lastly, Apple has a reactive approach when it comes to sustainability and CSR.

Dell, another major player in the PC industry and also US-based, signaled its commitment towards a new sustainability strategy in 2012. Its Dell 2020 Legacy of Good Plan brings the rest of that strategy into focus and sets the path for how social and environmental sustainability will become an accelerator development (Dell Website 2013.). Its long-term sustainability plan focuses on taking action to benefit the environment, strengthening its communities, and engaging its people in a diverse and inclusive workforce (Dell Website 2013). Ultimately, the notion of adding value in these areas flows directly from Dell's mission to enable people everywhere to grow and thrive and reach their full potential.

HP, another US multinational ICT firm, has fully embraced its social and environmental responsibilities through technology, employees and partnerships across institutions and industries (HP Global Citizenship Report 2012). HP emphasizes its efforts in corporate ethics, environment sustainability, product solutions, return and recycling, business operations, human rights, supply chain, people management, and social innovation. The HP Company Foundation coordinates efforts with HP global citizenship initiatives. It awards grants for humanitarian relief in communities hit by disaster, as well as for education, and philanthropic initiatives.

Such multinationals have played a key role in attracting foreign direct investment, generating employment, providing training and raising wages in the PC sector worldwide. Like Acer, Dell and HP, Lenovo also saw a need to somehow address social, labor and sustainability challenges arising from their operations. What sets Lenovo apart from these other multinational PC companies, however, is the social, political and economic setting of its home base, China.

CSR in China

China has the world's largest population. Its economy has also become the second largest and one of the fastest growing, with gross domestic product (GDP) expanding at a cumulative average growth rate of 10% over the past 30 years (IMF 2013). The International Monetary Fund expected China's economy to be bigger than that of the USA by 2014 (Yueh 2014).

However, China remains a developing country. Official data from the World Bank show that about 100 million people still lived below the national poverty line of RMB 2300 (US$ 375) per year at the end of 2012. With the second largest number of poor people in the world after India, poverty reduction remains a fundamental challenge in China (Luo and Zhu 2014).

Jiang Zemin, a former premier during the 1990s, was seen as promoting economic growth at nearly any cost. While economic growth has certainly benefited China, it also ignored significant social costs. One often cited example of that government priority and conscious trade-off is that 20 of the world's 30 worst polluted cities are in China. The urban-rural income gap has also widened significantly across China.

In 2006, the Chinese government and Chinese Communist Party under Hu Jintao, Jiang's successor, launched a comprehensive campaign to raise awareness and support for a series of specific policies and programs under the two labels Modern Socialist Countryside and Harmonious Society. According to Premier Wen Jiabao (Ahlers and Schubert 2009), "The issues concerning agriculture, rural areas and farmers are fundamental ones that have a bearing on China's overall modernization drive". A new "socialist countryside" would put agriculture and rural areas on the agenda of China's modernization campaign, with policies to engage other industries to support agriculture and farmers and have cities support rural areas.

The Modern Socialist Countryside represents an attempt to narrow the rural-urban gap through a range of activities (Ahlers and Schubert 2009):

1. plan economic and social development in urban and rural areas as a whole, and firmly promote construction of the new countryside,
2. boost modern agriculture to consolidate industrial support for the new countryside construction,
3. ensure sustained increases in farmers' incomes to lay a solid rural economic foundation,
4. increase infrastructure construction in rural areas to improve rural material conditions,
5. accelerate development of public services in the countryside and encourage new farmers,
6. deepen comprehensive rural reform to guarantee systematic protection for rural people,
7. improve democracy in rural areas and perfect rural management, and
8. enhance leadership and motivate all party members and the entire society to care, support, and participate in the construction of a new countryside.

Another trend that Hu and the Communist Party were responding to was a significant increase in social unrest, with 74,000 incidents of "internal unrest" reported in 2004, mostly in the rural areas. Their response has been to announce a set of social objectives and specific policies to create what the Chinese Communist Party called an "Harmonious Society" (Xinhua News Agency 2006). Like the Modern Socialist Countryside initiative, it is focused on the need to broaden the benefits of economic development and to address the sources of anger, fear and uncertainty that driving the recent social unrest. Those feelings swelled as farmers saw their incomes and opportunities fall further behind those in urban areas, and as local officials illegally assessed taxes and fees or seized their land for commercial development for both personal gain and to meet revenue gaps as central government support declined. Other developments affected those in both rural and urban areas, including an inadequate social welfare system to support pensioners, the unemployed and ill, endemic corruption, pollution and other environmental problems, and escalating crime and divorce rates. The targets in China's 12th 5-Year Plan, covering the years 2011–2015, identified specific social and environmental challenges (US Chamber of Commerce – Asia 2012):

- *Water scarcity and quality*. In 2005, around 300 million rural residents lacked access to safe drinking water (Ministry of Water Resources 2009).
- *Industrial pollution and greenhouse gas emissions*. In 2006, 43% of China's cities did not meet category 2 air-quality standards. A 2009 report claimed this had decreased to 17%, but also acknowledged that some cities continue to suffer from serious pollution (Ministry of Environmental Protection of the People's Republic of China 2009).
- *Labor conditions*. Labor conditions had improved in the past decades, but the issue of excessive overtime was proving particularly difficult to address.
- *Product safety*. Product and food safety scandals and recalls have led to deaths, illness, and injury, damaging the reputation of the China country-of-origin brand overseas and sparking domestic unrest.
- *Corruption*. Widespread corruption was one of the areas of strongest public concern, and the Communist Party came to see it as a critical obstacle to the country's social and economic development. China scores 3.6 on the Transparency International Corruption Perceptions Index (where 10 indicate maximum trust in government integrity).
- *Inequality and economic exclusion*. Although a small proportion of individuals had become extremely wealthy, the incomes of many citizens had not kept pace with economic growth over the past decade.

Within this broad context, the drivers of CSR in China had been shifting towards its impact on Chinese firms' global brand and business practices (US Chamber of Commerce's *Asia 2012 Report*). Domestic Chinese pressure was rising, and citizens were finding ways to express disapproval over perceived poor corporate practices. At the same time, some Chinese firms were raising the bar for competitors by themselves adopting international standards of responsibility while moving up the value chain.

The Chinese government has also been a significant force in the adoption of improved CSR practices (Brubaker 2012). It has embraced CSR as a tool to help address social and environmental issues accompanying the country's rapid but uneven economic transition. The government has increasingly looked at a broader role for the private sector in national development. Firms have come to be seen not only as key actors in areas of investment, training and technology development, and also as partners in helping to build the institutions, policies and local capacity needed for a sustainable economy. Recent indications of this include:

- The new Company Law of 2006 which states, "Corporations in their business operation must abide by the laws, regulation, social and business morality and good faith rules, must accept supervision by government and the public, and must undertake social responsibilities."
- In 2007, the Ministry of Commerce issued the *Circular on Enhancing Environmental Surveillance on Exporting Enterprises* to restrict socially irresponsible enterprises from conducting foreign trade (MOFCOM 2007).
- In 2008, the State-Owned Assets Supervision and Administration Commission under the State Council published the *Guidelines on CSR Fulfillment by Central-Level Enterprises* (SASAC 2008).

These and other developments reveal a particularly strong CSR focus in China on ensuring compliance on social and environmental issues (SASAC 2008).A 2011 study concluded that CSR in China shares characteristics with other Asian countries (Herrera and Roman 2011). In China, as in India, Indonesia, Vietnam and the Philippines, CSR is influenced by legislation and regulation. In particular, behavior concerning labor and the environment promotes responsible behavior. In these countries, multinational corporations (MNCs) implement their own global policies aligned with international standards and provide a benchmark for local corporations and often provide local governments and regulators with a comparison base. In all of these countries, however, the relative scarcity of technical expertise is a key challenge in implementing CSR. Still, there is increasing interest in integrating CSR into operations, and concerns of competitiveness in this area internationally and the need to work in a cooperative manner with host communities have become important drivers for CSR. Local leadership is also important in promoting corporate CSR practices in these contexts (Herrera and Roman 2011).Most CSR activities practiced in China were philanthropic activities and programs that addressed specific social and environmental footprint concerns in their areas of operations. These initiatives were typically responses to international competitive realities and the need to comply with local laws and regulations (Herrera and Roman 2011).

Herrera and Roman (2011) identified salient factors that affect decision-making at the operational level regarding CSR-related matters in China, including:

1. Global corporate policies require multinational companies operating in China to observe the same standards used in their home countries, in addition to compliance to local regulations.
2. Companies in China have begun to put in place mechanisms and structures that encourage employee participation in CSR programs, and some companies included CSR employee participation as an important component in the performance evaluation process.
3. Implementation of CSR initiatives and programs in China is handled by different departments or groups within the organization.

CSR implementation has brought improved relationships and engagement with local communities, employees and the government. It also enhanced corporate reputation and competitiveness in the global market. Companies in China have begun to measure the impact of their CSR programs and initiatives and are creating benchmarks anchored on international standards (Herrera and Roman 2011).

A summary of key influences on the practice of CSR in China is provided in Fig. 3.1.

The challenge for CSR practice in China is how to engage small and medium sized local companies who are unable to comply because of the significant costs involved in implementing CSR programs, especially programs that are geared towards compliance with international standards (Herrera and Roman 2011).

Moreover, the level of integration of CSR differs between multinational companies and local Chinese companies, which can be attributed to a difference in perspectives. Western companies do not consider compliance as part of CSR, while Chinese companies consider compliance as the basic element of CSR. It is usual for multinational companies to adhere to global corporate policies and international standards, which in many cases go beyond what is required by local laws. Many local companies, however, are not even willing or able to comply with local regulations (Herrera and Roman 2011).

Larger private firms, however, in order to enhance their corporate reputation and brand value, have been accelerating the diffusion of CSR practices in China that enhances their social capital vis-à-vis the government, communities and other stakeholders. Supply chains are also coming to be seen as another force that can encourage Chinese companies to adopt CSR practices. In particular, multinational corporations are responding to their own stakeholders to implement CSR practices across their value chains, and are driving the adoption of CSR practices in the Chinese firms that they work with (Herrera and Roman 2011).

In China, as well as in other Asian countries, there is increased interest in taking a strategic approach to CSR. Additionally, as a result of the strong influence of the state, the alignment between corporate efforts and state goals is particularly important in China. Multinationals operating in China must take into account the need to balance the state's goals with its global branding. Thus, strategic CSR in China has evolved to integrate business priorities with collaboration and strategic alignment to address societal concerns and national development priorities (Table 3.1).

The emergence of NGOs in China has been a direct result of the changes in both the Chinese state and society since the beginning of the reforms in 1978. Reforms have led not only to a relaxation of state control over the economy and society, but have also seen the state actively creating and sponsoring NGOs in order to transfer to them certain functions which it used to perform itself under the centrally planned system (The Economist 2014).

In the economic area, the government has sought to reduce its direct management role by establishing intermediary organizations, such as trade associations and chambers of commerce, to perform sectoral coordination and regulatory functions.

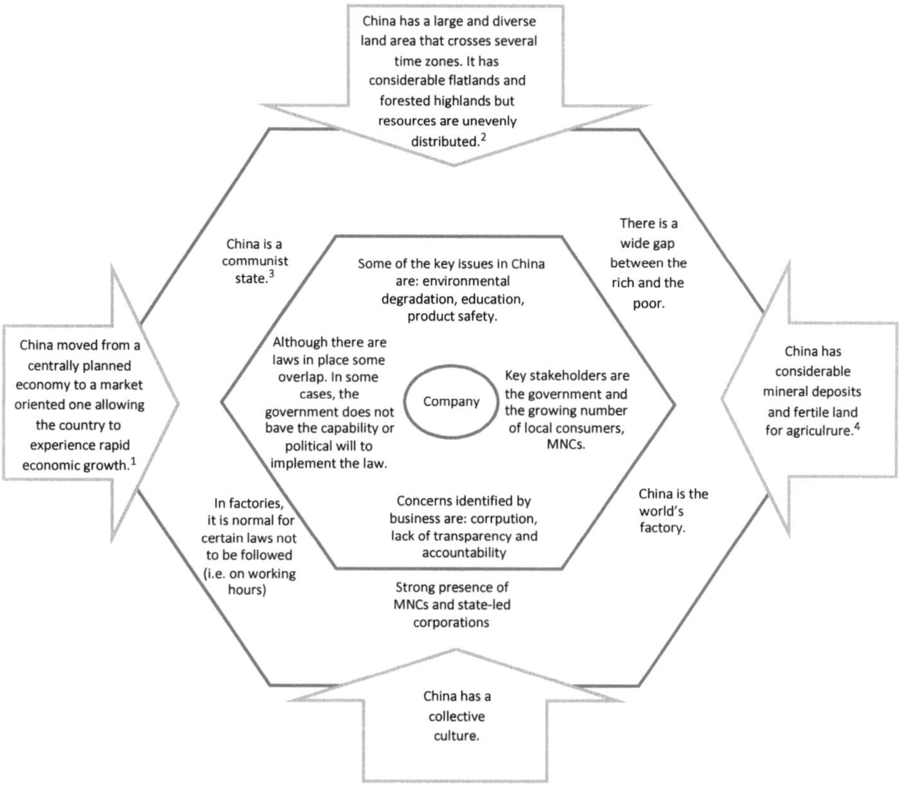

Fig. 3.1 Influences on CSR in China (Source: *Herrera, M.E. (*2011*). "Corporate Social Responsibility in Southeast Asia: An Eight Country Analysis," Ramon V. del Rosario Sr. Center for Corporate Social Responsibility)*

In the social welfare area, the government wants to foster an NGO sector to which it can offload some of the burden of service provision. The government also hopes that NGOs can mobilize societal resources directed towards social development to supplement its own spending (The Economist 2014).

NGO's have the potential to be an important player in furthering development in China, but they face many challenges in the Chinese environment. These include the absence of social policy to increase the capacity of NGOs, corruption, NGO dependence on the government, and China's political culture that is not conducive to civil society activism (The Economist 2014). Still, strengthening NGOs is broadly aligned with state's goals and represents a potentially effective means for companies to contribute towards China's development. The following section describes how Lenovo responded to this opportunity to support NGOs and at the same time contribute to the government's objectives for economic and social development.

Table 3.1 Recent trends and developments in four key areas of CSR

	Past CSR practice in China	Changes in progress	Emerging practice of CSR in China
Public policy	Viewed by government as a foreign-imposed requirement with protectionist aims.	CSR is viewed as a tool of public policy to encourage businesses to play a role in contributing to "scientific development" and "harmonious society".	CSR is becoming a key factor determining competitive advantage for companies in China and in the global economy.
Supply chain	Focused on ensuring supply chain labor and environmental standards through codes of conduct and audit	Tightened domestic labor and environmental laws and enforcement. Partnership and capacity building for labor and environment compliance. Recognition of the limits of audit approach.	CSR in manufacturing industry is focusing more on building capacity for environment health, safety and human resource management. Moreover strengthening the enforcement of environmental and labor laws.
International standards of practice and expectations	Concerned on many international standards initiatives that could represent trade and investment barriers intentionally or otherwise.	Careful endorsement and adoption of international standards. Aligning China's CSR framework to international standards.	Developing national standards of CSR practice to enhance the international CSR standards.
Internal and external stakeholders	Multinationals pushed for global CSR practices and signature programs, while state-owned enterprises have viewed CSR as a matter of upwards accountability to government.	Consumers, NGOs, media and other local stakeholder are expecting and demanding for responsible businesses.	Developing and testing bottom-up grievance and redress mechanisms to meet the demands of the growing consumers and civil society.

Prepared by the authors based on Simon Zadek et al. (2012). Corporate Responsibility and Sustainable Economic Development in China: Implications for Business. US Chamber of Commerce. https://www.uschamber.com/sites/default/files/legacy/international/files/17296_China%20Corp%20Social%20Responsibility_Opt.pdf. Accessed 06 May 2014

Lenovo's Corporate Development

Lenovo's beginning may be traced in Beijing in 1984, as the New Technology Development Company of the Institute of Computing Technology of the Chinese Academy of Sciences. The company is considered a pioneer of Chinese market reform in science and technology sectors. Initially it was founded not so much to create an internationally competitive technology enterprise, but to generate income from the information technology market to address the lack of government budget allocations. Up to that time, government budget allocations had been the sole funding for information, communication and technology as well as CSR activities in such organizations (Ahren and Zhou 2013).

The company's first technological innovation was derived from Institute and Computing Technology researcher Ni Guangnan, who developed a Chinese-language add-on card *(Lianxian)*, which was adopted as and remains the Chinese name of Lenovo to this day. In 1988 the company took on a small manufacturing firm, DAW, as a joint venture partner and added China Technology as a financial partner and expanded to Hong Kong. That same year the company was reorganized into Legend Computer Group.

During 1980s and 1990s, the Chinese government restricted foreign companies from establishing distribution and retail operations in China, so foreign companies had to use Chinese firms to sell and service their computers. Legend became the sales, distribution and service representative in China for a number of American and Japanese computer companies. This period was very beneficial for Legend, as it acquired both technical and management expertise from its Hong Kong operations and from its foreign partners, HP in particular (Ahren and Zhou 2013).

At that time, American and Japanese companies did not bother to bring the most technologically advanced and latest model PCs into China because of the low income of Chinese consumers. This provided an opening for Legend, with a strategy focused on localization and on the consumer market. In 1996 Legend slashed prices four times, just above cost, while introducing the latest central processing unit in its model. This price-based strategy worked. Legend saw a dramatic gain in market share and became the top PC seller in China, a rank it has maintained to this day. Legend continued to compete on price with the larger companies, executing a low-margin and low cost structure strategy to capture the the "middle" of the pyramid in the PC market (Ahren and Zhou 2013). By end of 1998, Legend's share of the Chinese market was 19.9% and was ranked third across the Asia-Pacific region (excluding Japan). By the early 2000s, Legend's market share reached 30%, and management began to look into international expansion.

Legend rebranded itself as Lenovo in 2003 and made a dramatic entry onto the world stage in 2004 with its U$1.75 billion acquisition of IBM's Personal Computing Division. That acquisition and transition from a China-centric to global company involved many challenges for Lenovo. Many observers doubted they would be able to successfully integrate two different cultures, languages, processes and markets (Spooner and Kanellos 2004). Those involved internally in the integration saw language and culture as two of the major challenges. Yang Yuanqing, Lenovo's CEO, recalls their initial global operations call, where most Lenovo executives could not understand everything that was said. Moreover, the Lenovo executives did not have a good understanding of Western culture. They were far from sure that Lenovo's success in China could be replicated globally (Kirkland and Orr 2013).

Ultimately, Lenovo proved successful in uniting what it describes as the best of Eastern and Western technology firms to establish a global personal computer brand.

It has expanded internationally by establishing roots in major markets and investing in sales, distribution, domestic manufacturing and R&D strategically. This has been Lenovo's implementation of its own version of a "global-local" company; i.e., building a local footprint either organically or through acquisitions in its major markets.

Lenovo has managed its international growth strategy successfully. Besides its acquisition of IBM's PC division, it formed a joint venture with NEC in 2011 to create the number 1 PC company in Japan. Later that year, Lenovo acquired Medion, a German consumer electronic company, for $630 million. In 2012, Lenovo formed a joint venture with EMC focused on servers and enterprise solutions, and also acquired Stoneware, a cloud software firm. Lenovo went on to acquire CCE, a leading consumer electronics company in Brazil, for $126 million in 2013. The next year, Lenovo bought IBM's low-end server unit for $2.3 billion and Motorola Mobility from Google for $2.91 billion.

Lenovo's main operating centers are located in Beijing, Paris, Raleigh, and Singapore. Lenovo also has subsidiaries in more than 60 countries, including R&D centers in Japan (Yamato), China (Beijing, Shanghai, Shenzhen) and the USA (Raleigh, North Carolina). The company sells its PC products in more than 160 countries and employs almost 21,300 worldwide (Lenovo Website 2014).

Yang explains how the company has gained even greater momentum in recent years. In the past, Lenovo was seen as a Chinese local brand and now other players in the industry are viewing the company as a serious competitor. *"I think some were too slow to address the mobile-internet trend. And one reason is that they believe you can outsource everything, not only manufacturing but even research and development. When you do that – when you rely on external parties to think for you – you lose the spark of innovation. Our momentum has been stronger than our competitors', but not because we quickly follow whatever Apple does. We don't. It's been stronger because we saw the market trends a decade ago and have invested in innovation."* (Kirkland and Orr 2013).

Yang's vision for the next 10 years is for Lenovo to become a strong global company across all of its segments and technologies, including PC and mobile, consumer and commercial, and mature and emerging. He hopes to transform the company from a PC market share leader into a "PC-plus" innovation leader. This would be the basis for sustained growth and profitability as well as provide the foundation for Lenovo to be a great global company that can last for generations.

Lenovo's Internal Context

This section provides a summary description of Lenovo's fundamental strategy and internal environment to provide further context for the VPP initiative.

Strategy Lenovo's vision is to create personal devices more people are inspired to own, a culture more people aspire to join and an enduring, trusted business that is well respected around the world. This vision guides the company in pursuit of its mission to become one of the world's great personal technology companies.

Structure Following its success in becoming the top PC firm globally in 2013, along with its acquisition of IBM's low-end server business and Motorola Mobile, Lenovo announced an organizational restructuring in January 2014. Lenovo established four new distinct business groups from its two original groups (consumer PC and mobile, and Think business group). The four new groups are: PC Business Group (includes Lenovo and Think brands), Mobile Business Group (includes smart phone, tablet, and smart TV), Enterprise (includes servers and storage), and Ecosystem and Cloud Services (includes both Android and Windows opportunities). All business groups are charged with continuous innovation, generating profits and expanding Lenovo's PC-plus business worldwide.

Lenovo's geographic sales operations were divided into America, Europe and Middle East Asia, China and Asia Pacific. The diverse Asia Pacific region was divided between Mature Markets (Japan, Australia and New Zealand) and Emerging Markets. Yang explained that the new structure was necessary to build new businesses and new pillars for the company. It would help Lenovo become faster, focused and more efficient in providing innovative products and services to diverse global markets.

Processes and Policies In terms of operations and policies, Lenovo and its business groups are for-profit operations are held accountable for meeting corporate targets. All business groups are also expected to subscribe to Lenovo's values and principles. Lenovo is implementing policies supporting the following specific priorities: sustainability, environmental responsibility, product safety and employee welfare, ethical business practices, and stakeholder engagement.

Values Lenovo sees its values as serving as the foundation of the company and defining its corporate identity. They are formally stated as: *Serving Customers, Trust and Integrity, Teamwork across Culture and above all these, Innovation and Entrepreneurial Spirit.*

In terms of innovation, Lenovo places more emphasis on incremental rather than groundbreaking innovation. It pursues the "right" technologies for the markets its competes in. Acquisitions, collaboration with industry associations and investments in research and development support this approach. While Lenovo, like many Chinese firms, are often criticized for being adaptive rather than innovative, this direction has worked well for Lenovo. It has managed to establish a firm competitive advantage based on meeting market trends and opportunities. Also, such incremental and adaptive innovation means that Lenovo has not made a costly miscalculation (Ahren and Zhou 2013).

Culture The company's culture has evolved significantly. At the time of Lenovo's acquisition of IBM's PC division in 2005, the culture of the company was very Chinese. Adapting it so that the IBM integration would be successful was the job of Steve Ward, who led management globally. Formerly the Chief Information Officer of IBM, Ward was appointed as the Chief Executive Officer of Lenovo. Globalization would enable Lenovo to potentially tap expertise from around the world. To facilitate this, the decision was made to make English as the common language within the company. This and a number of specific actions have resulted in a strong culture in which employees are encouraged to be innovative and entrepreneurial. This was formalized in a new set of core values whereby the company seeks to "continually improve and never be satisfied with anything less than a competitive advantage in technology, price, quality delivery, responsiveness, speed and efficiency" (Osawa and Luk 2014).

Its internal diversity is emerging as one of Lenovo's key strengths. With corporate roots in both the East and the West, Lenovo is a uniquely polycentric firm. It has leaders and an employee talent pool spread across the globe, who also shares a deep commitment to innovation and long-term responsibility. This drives its focus on anticipate and serve the complex needs of its customers and local stakeholders around the world.

Leadership Yang Yuanqing is credited with creating a performance-oriented culture instead of the traditional Chinese work style of "waiting to see what the emperor wants" (The Economist 2013). As the current Chairman and Chief Executive Officer of Lenovo, he continues to drive the message of innovation across all business operation globally. He has adopted a management style that allows employees freedom, shows confidence in personal capacity, and promotes based on performance.

These six drivers: strategy, structure, processes and policies, values, culture, and leadership form a governance frame for embedding and institutionalizing CSR and corporate social innovation into the business. The next section will describe the emergence and operation of Lenovo's discuss how Lenovo's Venture Philanthropy Program evolved.

Background to Lenovo's VPP Initiative

Attuned to China's challenges to finding solutions to problems of social and environmental sustainability without compromising national economic development goals, Lenovo's management decided that the firm could best contribute by leveraging its core competencies and partnerships. This led it to adopt the theme of creating a better world through information technology empowerment.

Lenovo's Corporate Sustainability Policy was implemented from 2008 in an effort to further formalize and integrate sustainability initiatives throughout the company's worldwide operations. All business units were expected to adhere and act according to Lenovo's sustainability principles across social, environmental and economic domains. Sustainability plays an integral role in Lenovo's "protect and attack" business strategy that includes establishing profit pools, pursuing profitable growth, expanding and solidifying China's leadership in smart phone and tablet markets, and building a stronger enterprise system. Lenovo has defined and continuously worked in six core sustainability focus areas: transparency, climate change, compliance and risk management, building sustainability culture, stakeholder relations and product leadership (Lenovo Sustainability Report 2013).

While implementing these principles throughout its business operations would certainly benefit society, there was a strong drive within Lenovo to undertake activities that would be seen as purely philanthropic and that would also benefit the company in a positive way. After reviewing a range of corporate philanthropy models, Lenovo settled on venture philanthropy.

Venture philanthropy is a strategic, focused approached to philanthropy through which funders provide substantive financial and in-kind support to help grantee organizations scale their work and achieve operational sustainability. Funders that take a venture philanthropy approach usually work with grantees for an extended period of time and play a more active role in shaping the future direction of grantee organizations (BSR 2012). Evidence from other corporations' venture philanthropy activities (e.g., Nike, Intel, Goldman Sachs and HP) suggest that five factors contribute to a strong venture philanthropy program: (1) clear focus, (2) impact measurement, (3) employee engagement, (4) leveraging the firm's sources of in-kind support, and (5) scalability.

In deciding the focus of its venture philanthropy program, Lenovo recognized the potential contribution of NGOs in promoting and implementing sustainable practices and policies in China (Campbell 2007). As the Chinese government was encouraging firms to create and embed social and environmental values into their activities, NGOs were able to refer to government regulations and policies when negotiating CSR-related issues with local and multinational firms. Although the government still closely monitors NGOs in China, these have become increasingly active in society, especially in matters related to the natural environment (Lin 2010) and consumer rights (Moon and Shen 2010).

The number of NGOs in China had grown significantly in the past decade. By 2009 there were more than 425,000 registered NGOs in China (Ministry of Civil Affairs 2010). This growth resulted from the government's inability to provide all the necessary social programs, some of which were consequently transferred to NGOs. Business for Social Responsibility Vice President, Jeremy Prepscius identified that what drove this development was the continued evolution of the Chinese society, which allowed civil discourse in China and was enabled by technology (Reeves 2012).

While the number of NGOs in China had been increasing, few had gained enough capacity to sustain their activities and make a significant, long-term impact. Most do not have working relationships with relevant ministries and agencies in the central government. They also typically lack adequate funding, technology and skills to maintain their relatively small-scale activities, much less expand.

Lenovo saw that it could use its resources and capabilities to help promising NGOs not only to achieve their own sustainability, but also to grow into significant contributors to China's national development. This also reduced Lenovo's potential reputation risk, as its relationships with NGOs and their beneficiaries would be indirect and separate from its business operations (Lei and Zhu 2010).

Design and Implementation of Lenovo's VPP

The VPP objective was to support NGOs in carrying out institutional development projects across China by providing substantive funding, technology, marketing expertise, and capacity building to Chinese NGOs. Lenovo officially launched its VPP on 18 December 2007 with an initial fund of RMB 2.5 million (US$350 thousand). The program's key innovation was to offer competitive funding and Lenovo's resources to promising small-to-medium size grassroots NGOs and social entrepreneurs addressing challenges in education and the digital divide, the environment, and disaster relief (Lenovo Website 2014). The program would provide successful applicants with funding, training, volunteers, opportunities for collaboration and marketing expertise, as well as space in an incubator dedicated to NGOs.

Partners Lenovo collaborated with several key partners in developing and implementing the program, including CSR Corporate Citizenship in Action, Non-Profit Incubator, BlueFocus, and area experts (Lenovo Website 2014). Corporate Citizenship in Action is a consulting institute devoted to helping corporations realize their CSR strategies and operating models, and it facilitates cooperation among firms, social welfare organizations and government. Lenovo contracted Corporate Citizenship in Action to conduct internal and external stakeholder surveys and formulate Lenovo's CSR framework and future plans (CSR Consulting 2013). Non-Profit Incubator is a Shanghai-based non-profit organization with a mission to advance social innovation and cultivate social entrepreneurship in China. It provides integrated support for start-up NGOs and social entrepreneurs (NPI 2014). Lenovo worked with these two organizations to develop and implement the VPP's strategic plan, with key parts being to supervise funds and to coordinate activities between Lenovo and its beneficiaries.

Non-Profit Incubator and Corporate Citizenship in Action, together with Lenovo volunteers, provided technical support and guidance to NGOs for organizational development and strategy making, including developing a clear mission statement and means of evaluating their progress. The Lenovo volunteers also helped to design and implement standardized processes across all areas of the NGO's operations, including finance and accounting, leadership, program management, organizational evaluation and digital information management (Lenovo Website 2014).

BlueFocus Communication Group was China's leading marketing services group. Lenovo partnered with BlueFocus to help NGOs establish their brands and attract more public partners and resources (BlueFocus 2008).

Lenovo's expert committee members were drawn from private and public organizations. They evaluate and select the applicants that will receive program resources.

Program Operation The program solicited applications from NGOs focused on the themes that Lenovo has targeted: education and the digital divide, the environment, and disaster relief. The process is outlined in Fig. 3.2.

1. *Application*: Interested NGOs submit their proposals to the program.
2. *Evaluation*: A rigorous *initial evaluation* and *final review* ensure the quality of the program, as only the most promising candidates were offered places in the incubator. Lenovo and Non-Profit Incubator's appraisal team jointly screen the organizations. Specialized assessment tools evaluate organizations in terms of their social impact, operational capability, organizational capability, and overall fit with VPP (Lei and Zhu 2010).
   - *Social impact*: The founder has a good understanding of social problems and a keen grasp of social needs.
   - *Operational capability*: The NGO has innovative, replicable and sustainable solutions to meet social needs.
   - *Organizational capability*: The founder and members have a clear sense of mission and strong commitment to charity, and the founder demonstrates strong leadership and accountability.

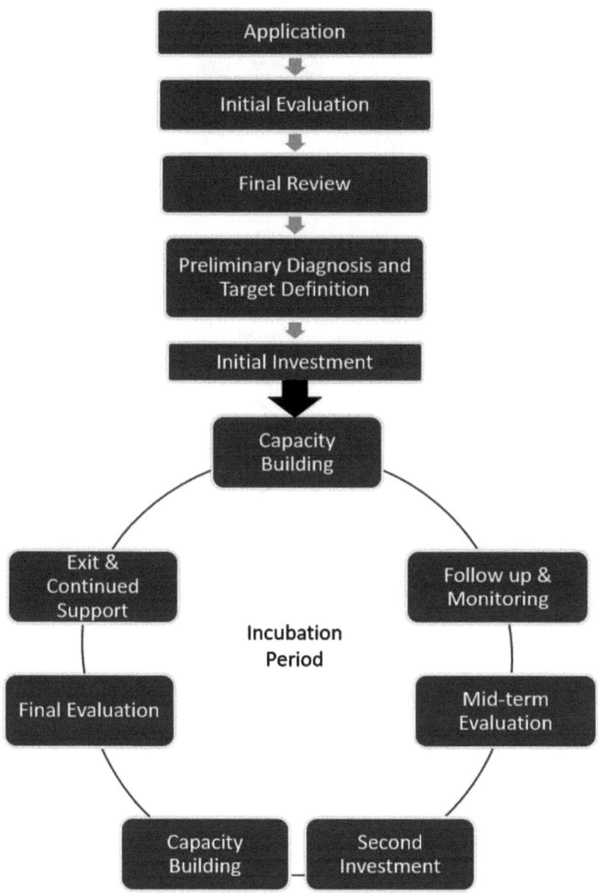

Fig. 3.2 Lenovo's venture philanthropy program incubation process

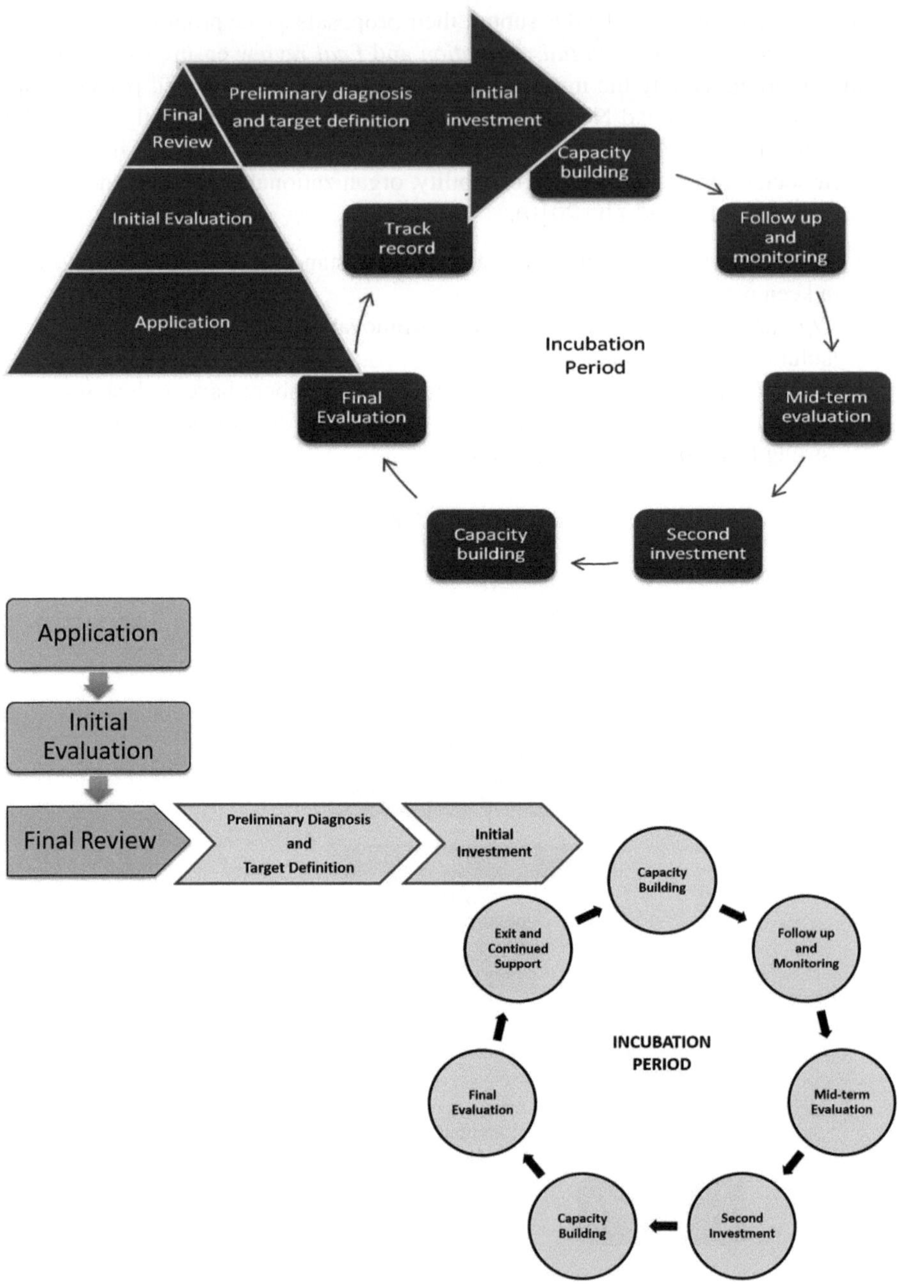

Fig. 3.2 (continued)

- *Fit with the Lenovo Venture Philanthropy Program*: The founder and members must demonstrate a strong willingness to work with a dedicated professional team from Lenovo's VPP.

3. *Incubation*: The incubation process begins with a preliminary diagnosis and goal-definition for the recipient NGOs. Based on this needs assessment, Lenovo provides resources and assistance related to office space, legal registration, operations and seed funding. Operational capability-building services include training in the areas of strategic planning, leadership, governance, program design, fundraising, human resource management, public relations, volunteer and program management, finance and accounting, and information management. Lenovo also facilitates its employees to volunteer in these incubation activities (Lei and Zhu 2010).
4. *Final Evaluation*: At the end of the specified incubation period, the NGO's growth, strategy and business plan are evaluated (Lei and Zhu 2010).
5. *Exit and Continued Support*: The NGO will leave the incubator once it has reached the target level of development. Even after leaving the incubator, however, the NGOs are provided follow-up services, including consultation, until a strong track record is established (Lei and Zhu 2010).

The first round of Lenovo's VPP was in full swing by May 2008, with the initial funding budget of RMB 2.5 million (US$350,000) allocated to 16 NGOs.

Program Evolution Lenovo considered the first round of the VPP to be a success, and increased funding for the 2009 round to RMB 3 million (US$450,000). For this second round, Lenovo decided to prioritize NGOs focused on environmental protection. They divided these funds between a subset of the first round (2008) grantees for which they continued support, and a target group of 10 new NGO recipients. By mid-2009, the new group of NGOs entered Non-Profit Incubator's Beijing-based incubator (Lenovo Website 2014).

Lenovo also modified the VPP to focus on youth, branding it the Youth Venture Philanthropy Program with the objective of encouraging college students to start their own businesses targeting public welfare and charity areas as a means of creating value for society. The program's goal was to find the best start-up teams from among college students and fresh graduates nationwide (Lenovo Sustainability Report 2009). Lenovo would offer its professional resources to assist in training, public relations, venture capital, and other skills to help prepare these selected students and fresh graduates and grow successful social enterprises.

Lenovo continued its focus on young social entrepreneurs in 2010 and launched the Lenovo Youth Social Entrepreneurship Contest as the third iteration of the VPP. The contest's theme was *Showing Your Love and Creating the Future*. Lenovo hoped that this would not only help identify and support promising young social entrepreneurs, but also spur more organizations to pay attention to youth-led charities and entrepreneurial activities. This, Lenovo felt, would bring innovation and broader, longer-term benefits to China's charity sector.

The contest was open to all youth groups, but was primarily marketed to college students. Lenovo encouraged any ideas or business plans following the contest's theme. Compared to the program in 2009, Lenovo engaged in more extensive outreach through media and events, including lectures on social entrepreneurship at 120 universities and colleges. About 100 experts and leaders from NGOs and social enterprises shared their knowledge and experiences in social entrepreneurship and philanthropic careers.

After screening more than 18,000 proposals, 27 teams from across the country were selected to enter the NBI incubator and receive RMB 10,000 (US$1500) in seed money (Yannan 2010). They were also eligible for other support, including professional training, internships at famous charity organizations, and volunteer assistance from Lenovo employees.

Lenovo continued its focus on youth-led social entrepreneurship under the VPP in 2011, this time choosing *Micro-charity can make a Big Difference* as the contest theme. Using the Chinese micro-blogging site SinaWeibo as its platform, Lenovo launched a micro-blog competition for young people to contribute to a virtual discussion about environmental protection, education and the digital divide, and community development. Contest contributors submitted descriptions of their activities for judges to evaluate and for others to comment on. Lenovo and its partners selected nearly 100 submissions and winning teams received support including seed money, training and other services to build the charitable groups' capacity (Yannan 2010).

In 2012, Lenovo shifted the program from youth-led social entrepreneurship to identify ten promising NGOs under the theme *Seeking Good Deed Do-ers*. RMB 1 million (US$1.6 million) in grants, in addition to in-kind support, was allocated for distribution to ten NGOs. The program explicitly integrated Lenovo's corporate mission–*Connecting the world with science and technology*—with the five areas that VPP had come to focus on; namely, narrowing the digital divide, environmental protection, education, community development and culture. Lenovo hoped to highlight the role of NGOs in leading developments in these areas. The new program also used social media extensively to promote project submissions and public awareness.

The NGO selection criteria (Yannan 2010) for the 2012 iteration were also revised and made more explicit:

- *Innovativeness*: "Out-of-the-box" thinking and adoption of information technology approaches that have not been widely implemented and the adoption of new strategies or initiatives that address social needs or concerns.
- *Feasibility*: Demonstrated practicality and sustainability of the project in addressing social needs or concerns.
- *Impact:* Scale and scope of the program's impact on the targeted social needs or concerns.
- *Operational and leadership strength*: Managerial ability to execute the program, sustains the organization, build stakeholder support, and provide examples of best practice.

Lenovo also introduced a mentoring process into the competition by providing personalized guidance and training through a 3-stage process. An initial group of 30 projects were selected, and then experts and opinion leaders submitted their comments through Sina Weibo. Based on these, 15 NGOs were chosen to receive in-depth guidance from advisors and support in the form of Lenovo products. In the final round, 10 NGOs were awarded RMB 100,000 in grant money and went through the incubation. These NGOs received assistance in the areas of training and capacity-building, and through resource specialists and employee volunteers that were provided by Lenovo (BSR 2012).

The first 5 years of the program has helped 65 NGOs, enabling them to scale their efforts across China in the areas of education, environmental protection, narrowing the digital divide, and poverty alleviation. Lenovo hired Business for Social Responsibility to evaluate each aspect of the VPP and provide recommendations that would be used to develop a new 5-Year Strategy.

Program Review The report by Business for Social Responsibility focused on three dimensions for improvement. The first included the program's basic strategy, goals and design. VPP had a relatively clear objective and target audience; namely, to support Chinese NGOs. However, Lenovo had not articulated a clear definition of "venture philanthropy" nor specific goals and indicators that could be used to assess the program's success or degree of impact. The study recommended that Lenovo develop clear indicators of impact in three dimensions: Lenovo's value-adding role, the NGOs' overall capability improvement, and the NGO's increased impact on its beneficiaries (BSR 2012).

A core feature of the program was Lenovo's leveraging of its resources—funding, technology, marketing, and publicity expertise—to help the NGOs. However Lenovo needed to be clearer about the purpose of the project and the appropriate use of a marketing strategy. Furthermore, the program design did not incorporate any monitoring or impact measurement, nor did it evaluate the sustainability of the chosen NGOs growth and development (BSR 2012).

The second dimension of the program includes its execution. The study focused on the implementation team and their training, the NGO selection and capability-building process, and publicity management. While the implementation team seconded from Lenovo and the partner organizations together undertook project management and controlled internal resources, the Lenovo team had no specific expertise in the area of social problems. Moreover, frequent turnover in the project team hurt the program's operational continuity (BSR 2012).

The NGO selection was successful, especially given the large numbers of applicants to the annual contests. Business for Social Responsibility felt that Lenovo had effectively addressed the concerns of perceived bias in voting and final selection of NGOs that had arisen in the early years of the program.

While the study assessed the training for selected NGOs to be well-designed and delivered, the need to provide more targeted advising to the NGOs given the wide variation in their needs was also highlighted. Some recommendations include that the program introduce NGO-specific training and mentoring based on a needs assessment of each NGO. Moreover, Lenovo should explore options on how to provide more customized assistance to grantees so that support can have the maximum impact for them and their beneficiaries.

The study was particularly critical of the third dimension, namely program monitoring and evaluation. It described reporting, feedback and follow-up with winning NGOs as poor. Lenovo did not have a systematic process for evaluating the impact of the program on NGOs. The study shows that Lenovo did not build longer-term relationships with the NGOs it supported.

The study made several recommendations for improvement. First, they proposed that Lenovo use goal-setting to drive alignment through the program and across the partners, and use measurement and reporting processes to inform continuous program improvement. Lenovo would have to set more specific long-term and short-term goals that link with the intended impact on the NGO's capacity, the company, and society more broadly. Complementing this, Lenovo should develop an ongoing reporting process and encourage NGO leaders to use that as feedback to reflect on their successes, challenges and opportunities, and to guide their own development.

They also cited the need for more efficient coordination of activities among partners and stakeholders by better clarifying roles and responsibilities and managing team members. Project partners such as Non-Profit Incubator and BlueFocus would have to more effectively and efficiently manage finances, external stakeholders and communication, as well as monitor progress.

Lenovo would also have to apply corporate resources more effectively to meet each NGO's particular needs. To do this, the program team must develop a needs-assessment process, adapt capability-building content to those needs, and introduce measures to assess whether those needs were met. Needs assessment could be designed into the initial application process and deepened in later evaluation stages.

Finally, Lenovo could do more to help the NGOs develop relationships with critical stakeholders (for example, key departments of the Chinese government), relevant international organizations, experts and other NGOs as part of their capacity-building. They could also have deeper interactions with experts during the selection stage. As part of the mentoring process for selected NGOs, Lenovo could introduce experts relevant to the NGO's specific focus. Lenovo could also facilitate more exchanges and collaboration among the NGOs it supports.

Business for Social Responsibility summarized their key recommendations as follows:

1. use goal-setting to drive alignment throughout a project across partners,
2. identify NGO's specific needs and adapt accordingly to meet them,
3. coordinate with partners and stakeholders with case to be efficient,
4. provide opportunities for beneficiaries to enhance their network,
5. develop useful and timely reporting mechanisms for continuous learning,
6. use corporate resources to meet specific needs, and
7. deploy additional resources and strategies to build NGO capacity.

Discussion

The case of Lenovo's VPP provides a number of insights and dilemmas that firms must face when developing CSR strategies for a global firm and brand. In terms of implementing such a program, Lenovo's CSR manager underlined the challenge that diversity of the social sector and created for selecting among NGOs. The level of the NGO's social impact and the character of the long-term effect of Lenovo's support had to be assessed. Long-term sustainability of NGO's that graduated from the program was also a major concern. Specifically, would the support provided by Lenovo to the NGOs foster dependence and, ironically, work against the program's goal of improving long-term sustainability?

Overall, Lenovo had unique competencies amongst its staff and in its technology, but those resources were not fully leveraged during the first 5 years of the program. More volunteers could have been offered, and more customized technology solutions could have been developed based on better needs identification and goal setting. While Lenovo did use its marketing resources to promote the NGOs rather than just Lenovo itself, it could have done more by helping NGO-specific campaigns (BSR 2012).

Lenovo saw its VPP as a linked to its corporate brand and mission, and also as differentiating Lenovo from its competitors. Unlike many companies who choose to support CSR activities and organizations unrelated to their businesses, Lenovo saw its CSR activities as well aligned with its overall brand strategy in China. This did generate benefits for the company.

However, an emerging question within Lenovo was how the VPP in China would be related to Lenovo's global CSR and branding strategy. Would it remain a China-based initiative? Could it, or should it, be expanded globally? This case illustrates the complexity faced by firms with global brands who must develop and implement CSR initiatives, that are often locally embedded, within the context and considerations of their global presence.

This highlights the importance, but also the dilemma, of balancing the realities of internal and external business considerations for firms based in a country like China. Successful global companies seek to balance internal circumstances with global strategy, structure, branding and policy. For a global brand based in China, however, also balancing the state's expectations with the firm's internal and broader global context increases the challenge.

References

Acer Website. 2014. Corporate Social Responsibility. http://us.acer.com/ac/en/US/content/csr. Accessed 26 Apr 2014.

Ahlers, A., and G. Schubert. 2009. "Building a new socialist countryside" – Only a political slogan? *Journal of Current Chinese Affairs* 38(4): 35–62.

Ahren, N., and Y. Zhou. 2013. China's competitiveness. Case Study: Lenovo. Center for Strategic and International Studies. http://csis.org/files/publication/130129_competitiveness_Lenovo_casestudy_Web.pdf. Accessed 24 Apr 2014.

Barne, G. 2012. Ten grave problems facing China. The China story. http://www.thechinastory.org/2012/09/the-ten-grave-problems-facing-china. Accessed 24 Jan 2015.

BlueFocus. 2008. http://www.bluefocusgroup.com/en/about/jieshao. Accessed 26 Apr 2014.

Brubaker, R. 2012. China and sustainability: Connecting the dots between economy and ecology. http://www.theguardian.com/sustainable-business/blog/china-sustainability-economy-environment-ecology. Accessed 5 Apr 2014.

Business for Social Responsibility (BSR). 2012. Lenovo's venture philanthropy approach supports Chinese NGOs. http://ciyuan.bsr.org/data/resources/ciyuan-lenovo-venture-philanthropy-en.pdf. Accessed 22 Feb 2014.

Campbell, J.L. 2007. Why would corporations behave in socially responsible ways? An institutional theory of corporate social responsibility. *Academy of Management Review* 32(3): 946–967.

CSR Consulting. 2013. Corporate citizenship in action. http://www.csrconsulting.com.cn/en/new1.htm. Accessed 26 Apr 2014.

Dell Website. 2013. Dell 2020 legacy of Good Plan Report. http://i.dell.com/sites/doccontent/corporate/corp-comm/en/Documents/2020-plan.pdf. Accessed 27 Apr 2014.

Herrera, M.E.B., and F. Roman. 2011. *Exploring strategic CSR: Sustaining and strengthening corporate social responsibility*. Manila: Asian Institute of Management Ramon V. del Rosario Center for Corporate Social Responsibility.

HP Global Citizenship Report. 2012.. http://h20195.www2.hp.com/V2/GetPDF.aspx/c03742442.pdf. Accessed 26 Apr 2014.

International Monetary Fund (IMF). 2013. World Economic Outlook. http://www.imf.org/external/pubs/ft/weo/2013/01/weodata/weorept.aspx?sy=2012&ey=2018&scsm=1&ssd=1&sort=country&ds=.&br=1&c=924&s=NGDPD%2CNGDPDPC%2CPPPGDP%2CPPPPC%2CGGXWDG_NGDP&grp=0&a=&pr.x=41&pr.y=17. Accessed 26 Apr 2014.

Kirkland, R., and G. Orr. 2013. Thriving in a 'PC-plus' world: An interview with Lenovo CEO Yang Yuanqing. McKinsey Quarterly. http://www.mckinsey.com/insights/high_tech_telecoms_internet/thriving_in_a_pc-plus_world. Accessed 12 Mar 2014.

Lei, J., and S. Zhu. 2010. The practice of social entrepreneurship as a model: Case study between Sweden and China. University of Gavle. http://www.diva-portal.org/smash/get/diva2:326217/FULLTEXT01.pdf. Accessed 04 May 2014.

Lenovo Sustainability Report. 2009. Built to last: Sustainability for the new world. http://www.lenovo.com/social_responsibility/us/en/FY2009_Lenovo_Sustainability_Report.pdf. Accessed 22 Feb 2014.

Lenovo Sustainability Report. 2013. A new era of sustainability: Lenovo's PC+ transformation FY 2012/2013 Global Sustainability Report. http://www.lenovo.com/social_responsibility/us/en/FY2013_Lenovo_Sustainability_Report.pdf#page=1&view=fit. Accessed 04 Mar 2014.

Lenovo Website. 2014. http://www.lenovo.com. Accessed 22 Feb 2014.

Lin, L.W. 2010. Corporate social responsibility in china: Window dressing or structural change. *Berkeley Journal of International Law* 28: 64.

Luo, X., and N. Zhu. 2014. What drives the volatility of firm level productivity in China?. Policy research working paper; no. WPS 6846. Washington, DC: World Bank Group. http://documents.worldbank.org/curated/en/2014/04/19447081/drives-volatility-firm-level-productivity--china-drives-volatility-firm-level-productivity-china. Accessed 26 Apr 2014.

Ministry of Civil Affairs. 2010. Statistics on the development of civil affairs. www.maca.gov.cn. Accessed 26 Apr 2014.

Ministry of Environmental Protection of the People's Republic of China. 2009. Report on the State of the Environment in China. http://english.mep.gov.cn/down_load/Documents/201104/P020110411532104009882.pdf. Accessed 26 Apr 2014.

Ministry of Water Resources. 2009. http://www.mwr.gov.cn/english/sdw.html. Accessed 05 May 2014.

MOFCOM. 2007. Commercial Information Circular No. 392/2007: Strengthening the Supervision on Export Enterprises (2007). http://www.tid.gov.hk/textonly/english/aboutus/tradecircular/cic/asia/2007/ci3922007.html. Accessed 27 Apr 2014.

Moon, J., and X. Shen. 2010. CSR in China research: Salience, focus and nature. *Journal of Business Ethics* 94(4): 613–629.

Non-Profit Incubator (NPI). 2014. http://www.npi.org.cn/index_en.html. Accessed 26 Apr 2014.

Osawa, J., and L. Luk. 2014. How Lenovo built a Chinese Tech Giant. *The Wall Street Journal*. http://online.wsj.com/news/articles/SB10001424052702303973704579352263128996836. Accessed 26 Apr 2014.

Reeves, J. 2012.The CSR challenge for companies doing business in China. www.forbes.com. Accessed 26 Apr 2014.

SASAC. 2008. Guidelines on CSR fulfillment by Central-Level Enterprises. http://www.sasac.gov.cn/n2963340/n2964712/4891623.html. Accessed 27 Apr 2014.

Schmitz, J. 2006. The implications of China's growth for other Asian countries. http://www.dell.com/learn/us/en/uscorp1/2020-goals?c=us&l=en&s=corp&cs=uscorp1. Accessed 02 Apr 201

Spooner, J., and M. Kanellos. 2004. IBM sells PC group to Lenovo. CNET. http://news.cnet.com/IBM-sells-PC-group-to-Lenovo/2100-1042_3-5482284.html?tag=mncol. Accessed 02 Apr 2014.

Standberg, C. 2013. 7 reasons why CSR programs are good for industry associations. GreenBiz Website. http://www.greenbiz.com/blog/2013/10/15/7-reasons-why-csr-programs-are-good-industry-associations. Accessed 02 Apr 2014.

The Economist. 2013. From guard shack to global giant: How did Lenovo become the world's biggest computer company? http://www.economist.com/news/business/21569398-how-did-lenovo-become-worlds-biggest-computer-company-guard-shack-global-giant. Accessed 26 Apr 2014.

The Economist. 2014. Beneath the glacier. http://www.economist.com/news/china/21600747-spite-political-clampdown-flourishing-civil-society-taking-hold-beneath-glacier. Accessed 15 May 2014.

Unruch, G. 2012. Apple and the "Little Dutch Boy" strategy. HBR Blog. http://blogs.hbr.org/2012/05/apple-and-the-little-dutch-boy/. Accessed 26 Apr 2014.

US Chamber of Commerce – Asia. 2012. Corporate responsibility and sustainable economic development in China: Implications for business. https://www.uschamber.com/sites/default/files/legacy/international/files/17296_China%20Corp%20Social%20Responsibility_Opt.pdf. Accessed 24 Apr 2014.

Xinhua News Agency. 2006. China publishes 'Harmonious Society' resolution. http://www.china.org.cn/english/2006/Oct/184810.htm. Accessed 12 Sept 2014.

Yannan, T. 2010. Innovative students milk idea for all its worth. China Daily. http://www.chinadaily.com.cn/cndy/2010-09/27/content_11350744.htm. Accessed 06 May 2014.

Yueh, L. 2014. Is China about to overtake US as the world's largest economy? BBC. http://www.bbc.com/news/business-27216705. Accessed 5 May 2014.

Zadek, S. et al. 2012. Corporate responsibility and sustainable economic development in China: Implications for business. US Chamber of Commerce. https://www.uschamber.com/sites/default/files/legacy/international/files/17296_China%20Corp%20Social%20Responsibility_Opt.pdf. Accessed 06 May 2014.

Chapter 4
From Corporate Social Responsibility to Disruptive Innovation: Samsung's Green Memory Initiative

Poonacha K. Medappa and Shirish C. Srivastava

Section 1: Setting the Stage

> *Never before has there been such pressure on companies to address social and environmental responsibilities, and never before has there been such a wealth of opportunity to be derived from doing so – Lee and Kim* 2009

Corporate Social Responsibility (CSR) and Its Increasing Relevance

The definition of corporate social responsibility (CSR) has evolved over the years to refine and expand the scope and dimensions of the CSR activities. One of the most common concepts of CSR is the triad of economic, environmental, and social responsibilities, which was introduced by Elkington in his book "Cannibals with Forks: The Triple Bottom Line of twenty-first century Business, 1998". There have been some additions to this triad with new dimensions being included over time. In the business and society literature, it has been argued that specific dimensions of CSR include safety (Wokutch 1992), human rights (Jennings and Entine 1999), the environment (Fryxell and Dooley 1997), philanthropic donations (Wokutch and Mallot 1998), and local community considerations (Mallot 1998). With the concept of CSR more or less established, the focus of debate has now shifted towards its implementation and the resulting benefits (Lee and Kim 2009). What originally started as a responsibility

P.K. Medappa (✉) • S.C. Srivastava
Information Systems & Operations Management (ISOM) Department, HEC, Paris, Jouy en Josas Cedex 78351, France
e-mail: poonacha.medappa@hec.edu; srivastava@hec.edu

G. Lenssen et al. (eds.), *The Role of Corporate Sustainability in Asian Development*, Advances in Business Ethics Research 7,
DOI 10.1007/978-3-319-45160-2_4

Table 4.1 Top five drivers for CSR investment

What is driving business in your industry to implement more socially and environmentally responsible business practices?	
Top five drivers for CSR investment	Percentage of companies that indicated the corresponding driver (2014)
Cost management	67%
Client/customer demand	64%
Because it's the "right thing to do"	62%
Public attitudes/building brands	59%
Recruitment/retention of staff	58%

Source Grant Thornton IBR (2014)

that the companies adhered to; CSR has now evolved into value-adding initiatives that the companies are actively pursuing. Value addition through CSR occurs not just in terms of improvements to the society and environment but also in terms of enhanced reputation for the company and reduced cost for both the company and the end customers.

A recent report by Grant Thornton International based on interview data from 2500 respondents from 34 economies examines the motives of companies for undertaking CSR initiatives (Grant Thornton International 2014). The report identifies the potential for cost saving as the most important driver for engaging in CSR initiatives. The study suggests that firms are looking for more tangible benefits from their CSR programs as can be seen from Table 4.1 which highlights the top five drivers for CSR investments.

While there has been a global trend towards appreciating the cost saving/value adding opportunities meted by CSR investments, there have also been some variations in terms of approaches adopted to implement CSR activities by different industries. The electronics industry's approach to CSR is of particular interest to this case study and is detailed in the next sub section.

The Unique Case of Electronics Industry

In the case of the electronics industry, there is high relevance for pursuing sound CSR policies because of the industry's two distinguishing features (1) use of rare minerals mined from across the world including conflict zones resulting in strong environmental and political concerns; and (2) a complex production process that is spread across several suppliers and geographies resulting in concerns for CSR compliance across the value chain.

The electronics industry has recognized the need for a common voice with which the industry can set CSR standards and address global concerns. It is for this very purpose that the Electronic Industry Citizenship Coalition (EICC) was introduced in October 2004. In fact, the origin of the EICC can be traced to the time when the British NGO Catholic Agency for Overseas Development (CAFOD) pointed to the poor employee working conditions at the suppliers for various global brands such as HP, Dell, and IBM. Some of these companies (eight to begin with) came to agree that they needed to respond in a single voice to CSR concerns rather than to act independently resulting in the formation of the EICC (Lee and Kim 2009). Today the EICC comprises more than 100 electronics companies with combined annual revenue of approximately $3 trillion, directly employing over 5.5 million people. This group supports a common "Code of Conduct" for electronics companies – the EICC Code. The EICC code was created in order to establish standards to ensure that the working conditions in the electronics industry supply chain are safe, that workers are treated with respect and dignity, and that business operations are environmentally responsible and conducted ethically. Consequently, the code covers expectations for performance across a range of company practices, including labor, health & safety, environmental practices, ethics and management systems and has become the de-facto industry standard for social responsibility in the electronics sector. In addition to EICC members, their tier 1 vendors are required to implement the EICC Code of Conduct.

Since its founding, EICC has targeted several issues pertaining to sustainability across the electronics industry's supply chain. One of the key areas of concern for EICC is Green House Gas (GHG) reduction. According to a McKinsey Quarterly report published in 2008, the electronic industry is estimated to contribute to more than two percent of global GHG emissions. While this sounds like a small contribution, it is substantial – considering that the industry's emissions are projected to double by 2020. GHG emissions primarily stem from energy consumed during manufacturing and product usage. To address this concern, the EICC Environmental Reporting Initiative was created in 2009 to standardize the approaches to measure and report key carbon, water and waste indicators in the global electronics supply chain. The Environmental Reporting Initiative is based on global standards such as the Greenhouse Gas Protocol and established reporting formats such as the Carbon Disclosure Project (CDP) and the Global Reporting Initiative. One of the key benefits of this initiative is that it allows the companies to monitor their GHG emissions across its supply chain and also helps the companies set targets for reducing emissions.

Since their joining the EICC in 2007, Samsung Electronics has proactively participated in various initiatives of the EICC including joint actions on the industry's common issues. During the 2012 EICC Winter Membership Meeting hosted by Samsung Electronics in Korea, the company shared Korean CSR accomplishments and the local government's "shared growth" policies. The meeting also served as a valuable opportunity for the electronics industry to share CSR best practices with audiences in Korea. Samsung Electronics' commitment to EICC objectives is made evident by several initiatives that it has undertaken over the years. Specifically, its green management initiative that started in 2009 went beyond what EICC stipulated for its members by establishing a concrete set of goals and milestones to be achieved by the company in the areas of environmental sustainability.

CSR at Samsung Electronics – Green Management Initiative

From its beginnings as a small trading company in Korea; Samsung group has evolved to a world-class corporation with businesses that span advanced technologies, semiconductors, construction, petrochemicals, fashion, medicine, finance, hotels and more. At the end of 2013, the group operated out of 673 global offices employing over 489,000 people worldwide. Samsung Electronics is the flagship subsidiary of the Samsung Group, accounting for about 70% of the group's revenue. Since its inception in 1969, Samsung Electronics has grown into a global information technology leader. The company's offerings include home appliances such as TVs, monitors, printers, refrigerators, and washing machines as well as key mobile telecommunications products such as smartphones and tablet PCs. Samsung Electronics has also become a trusted provider of key electronic components like DRAM and non-memory semiconductors. Samsung Electronics operates in 220 locations around the world. At the end of 2013, Samsung Electronics' total employment stood at 286,284 with the domestic and the overseas workforce making up 33% and 67%, respectively.

Ceaseless Innovation has been central to Samsung Electronics' growth story. The focus on innovation can be seen as early as 1996, the year Samsung declared as the "Year of Design Innovation". Since then, the company has shifted its focus from sheer manufacturing volume to competing through better quality and design. The investment in R&D has seen steady growth from KRW 7.1 trillion ($6.15 billion), or 5.9% of its consolidated sales in 2008, to KRW 14.8 trillion ($12.82 Billion), or 6.5% of its sales, in 2013. While ceaseless innovation continues to drive Samsung Electronics to new heights, the company has also ensured that the responsibility it shares with the society and environment is not forgotten.

Samsung Electronics was one of the early proponents of EICC, joining the coalition in December 2007. In 2009, EICC set out to measure GHG emissions and energy consumption in the electronics supply chain. That year also saw Samsung Electronics making large strides towards CSR and green management. As a first step at the company, a CSR Liaison Office was set up under the CEO's supervision in January 2009. This office aimed to build a company-wide consensus and organize its corporate social responsibility practices. With the launch of the CSR Liaison Office, Samsung Electronics devised a company-wide CSR mid-term strategy, centralized its CSR communication channels, and established CSR meeting programs. In July 2009, the company announced the "Green Management Initiative" which created an ambitious set of goals and milestones. As part of this initiative, Samsung Electronics planned to invest KRW5.4 trillion ($4.68 billion) to curb greenhouse gases (GHG) and develop eco-products through 2013. The core activities of this initiative would include reduction of GHG generated from its business premises and product usage, expand the lineup of green products, investment in green R&D, foster a green workplace, and cooperate with business partners for green management.

A unique aspect of their green management plan was the development of eco-products, which intersected two important philosophies of Samsung Electronics – ceaseless innovation and "PlanetFirst" philosophy. This case study revolves around Samsung Electronics initiative of eco-product development and the subsequent creation of the green memory. The green memory story provides an excellent illustration of how commitment to the cause of environmental sustainability can result in the creation of an innovative product that not only provides competitive advantage to the innovator but also show signs of a disruptive "green" innovation for the industry.

Section 2: Samsung's Green Memory

"Our Green Management philosophy, which prioritizes the future health of our people and our planet and mandates environmentally responsible practices throughout our operations, drives our environmental sustainability initiatives. Let me give you one example: As part of our eco-design process, we consider the environmental impact of new products during their whole development cycle. We implemented this philosophy in 2009, and since then Samsung has invested 6.6 trillion KRW (6.3 billion USD) in sustainability efforts – resulting in a 50 percent reduction in greenhouse gas emissions and over 3,300 third-party verified product environmental certifications." – Oh-Hyun Kwon, Vice chairman and CEO of Samsung Electronics, Samsung Electronics Sustainability report 2014

Green Management Vision and Framework

On July 20, 2009, Samsung Electronics held a ceremony at its premises in Seocho-dong, Seoul, to unveil its green management vision and strategy. The underlying philosophy of green management was to contribute to building a prosperous society and preserving the environment through business activities that give due respect to people and nature. Based on this philosophy, Samsung Electronics established its vision for green management – "creating new value through eco-innovation" under the slogan of "PlanetFirst." Under the green management plan, a mid-term objective was established titled "Eco-Management 2013 (EM2013)". The EM2013 covered two major focal points (1) Reduction of greenhouse gas emissions relative to sales by 50% from the 2008 levels, (2) Development of energy efficient products and increase the usage of eco-friendly and recycled materials. To reach the stated goals, the company identified specific action plans under the three categories – green operation, green product, and green communication (Fig. 4.1). The category "green product" included investment in developing products that had high-energy efficiency. One of the results of the "green product" action plan was the

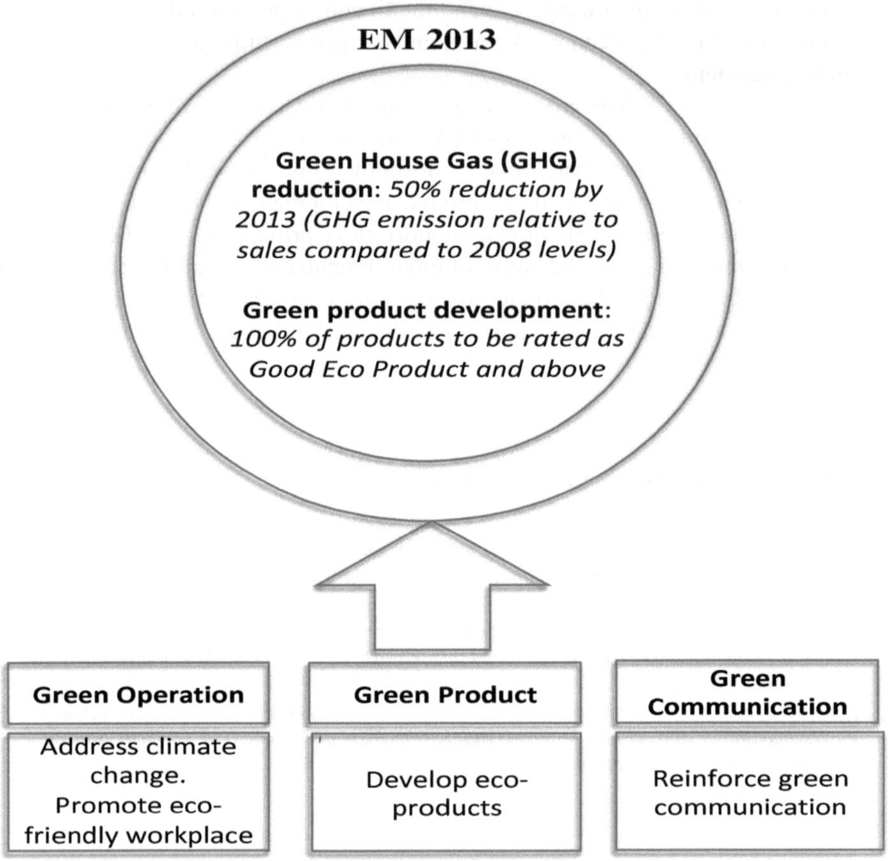

Fig. 4.1 Samsung Electronics eco management action plan released in 2009 (Source: Samsung Electronics Sustainability Report 2009–2010)

development of a new breed of energy efficient memory solutions that has been christened "green memory".

Eco-Product Development

One of the two key objectives of the EM2013 was eco-product development. To achieve this objective, Samsung Electronics established an "Eco-Design Process" that makes it mandatory to evaluate the eco-friendliness of a new product in the development stage. Based on these practices, the company developed an "Eco-Product Rating" process to evaluate the eco-friendliness of new products, accelerating the company's efforts to develop eco-conscious products and reduce waste. The highest eco product rating is that of "Premium Eco-Product" which strives to exceed

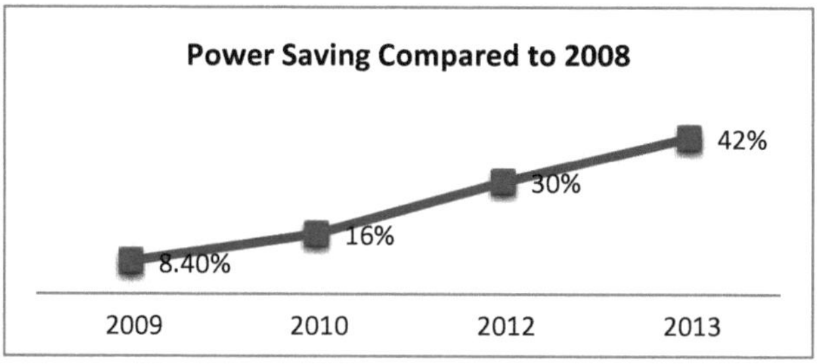

Fig. 4.2 Average reduction of power consumption of Samsung Electronics top consumer products (compared to 2008) (Source Samsung Electronics sustainability report 2013–2014)

the highest environmental standards and also provide innovative eco-conscious features.

An analysis done by Samsung Electronics identified greenhouse gas emissions associated with the use of its products was greater than its manufacturing processes. Thus, in order to achieve the EM2013 goal of reducing GHG emissions by 50% of the 2008 value, Samsung Electronics needed to increase energy efficiency in the product use phase as well. In this line, Samsung Electronics has managed to reduce the annual power consumption of its major consumer electronics products including TVs, refrigerators, and washing machines by 42% between 2008 and 2013 (Fig. 4.2). As a result of these efforts, GHG emission reduction to the tune of 88.6 million tons from 2009 to 2013 was observed (As cited in the Samsung Electronics Sustainability Report 2014). One innovation that has allowed Samsung Electronics to achieve this drastic reduction of energy consumption has been its green memory initiative. Green memory is the range of Double Data Rate (DDR) Dynamic Random Access Memory (DRAM) and Solid State Drive (SSD) memory devices that can increase system performance while reducing the energy consumption. These memory modules are integral for almost all electronic devices manufactured, specifically PCs, consumer electronic products and datacenters.

Green Memory as an Outcome of the Green Management Initiative

Since its entry into the Green Memory marketplace in 2009, Samsung Electronics has concentrated much of its R&D resources in low-power design technology. This effort was rewarded in 2013, when the company presented its 5th-generation green memory solutions. This solution incorporated several energy saving innovations resulting in state of the art Dynamic Access Random Access Memory (DRAM) and Solid State Drive (SSD) products. The 5th generation green memory solutions for the first time satisfied customers' expectations around performance, system installation space and energy consumption and simultaneously provided a clear return on investment. The innovations that has led to the 5th generation of green memory solution have been incremental that started way back in 2009.

In 2009, Samsung first began producing its 1st generation of green memory solutions based on the new DDR3 DRAM technology. The DDR3 proved to be a significant upgrade from the previous DDR2 DRAMs; exhibiting 73% energy savings (memory). In 2010, Samsung Electronics' commitment to the green management initiative led to its receiving six Eco-Design awards at CES 2011 (International Consumer Electronic Show), which was the maximum it had ever received. Amongst the products that won the award was their 2nd generation green DDR3 DRAM which qualified as the industry's highest-density memory module, designed for use in advanced server systems, providing high-density and high-performance with low-power consumption. This memory module could provide 83% power savings when compared to widely used DDR2 memory modules. The new DDR3 memory employed 35 nm fabrication process (as compared to the previous 46 nm process) and GHG reduction facility to achieve 24% reduction in GHG emissions and 14% reduction in power consumption. In 2011, Samsung introduced its SSD solution, which allowed it to leverage its innovation in DDR3 and package it as a complete memory solution. This SSD solution acted as an energy efficient, higher performing but costlier replacement to the more popular Hard Disk Drives (HDD) in datacenters and PCs. 2012 saw the emergence of the 4th generation of memory products, which provided upgrade to both its DRAM and SSD solutions. The key improvements were made in terms of idle power consumption and the employment of more efficient 20 nm fabrication process. The fifth generation of memory products, launched in 2013, saw a shift towards the new DDR4 from DDR3 DRAM and PCIe SSD from SATA SSD. According the Samsung Sustainability Report 2014, the key changes in the 5th generation memory solution were:

1. The "green DDR4 solution" simultaneously delivered high-speed, low power consumption, and high reliability
2. The "green PCIe SSD solution" delivered high efficiency with speed six times faster than SATA SSD

What it means is – a server mounted with the 5th generation DDR4 and PCIe SSD can enhance performance by more than 1.6 times compared to an old HDD server system. It can also expand system storage capacity four-fold by applying technology that eliminates redundant data, not only cutting power consumption and investment costs, but also minimizing installation space. As mentioned in the report, if all servers in the world were to be replaced with Samsung Electronics' "5th generation green memory solutions", the overall effect would be equivalent to a power saving of 45 Terawatt (TW) annually and the environmental impact would be comparable to planting 800 million ten-year-old trees.

Samsung continued to push the envelope of DRAM scaling with the production of its still popular DDR3 DRAM. In 2014, the company rolled-out the industry's most advanced 20-nanometer (nm) 4-gigabit (Gb) DDR3 DRAM. With the new 20 nm DDR3 DRAM, Samsung improved its manufacturing productivity. In addition, the new 20 nm 4Gb DDR3 based modules could save up to 25% of the energy consumed by equivalent modules fabricated using the previous 25-nanometer process technology. This improvement provided the basis for delivering the industry's most advanced green IT solutions to global companies.

The story of the evolution of Samsung Electronics' memory products from 2009 to the current date (Fig. 4.3) is a testament of their commitment to ceaseless innovation. Within each innovation that was observed in their memory products there is an underlying message of their "PlanetFirst" philosophy. Another striking feature is how disruptive the innovations tend to be. The rapidly evolving memory and fabrication technology resulted in potentially disruptive innovations emerging in the memory industry. These innovations tend to make the older products and production processes obsolete. Samsung Electronics' response to the disruptive nature of the innovation in the memory industry was the creation of new generations of memory products starting from 2009 (First Generation) all the way till 2013 (Fifth Generation) and beyond. Across these years Samsung Electronics looked at adopting technology, which allowed it to improve its energy efficiency and develop green memory products. In a press release in March 2014, Young-Hyun Jun, executive vice president, Samsung Electronics reaffirmed the company's commitment to the philosophies of planet first and ceaseless innovation –

> *Samsung's new energy-efficient 20-nanometer DDR3 DRAM will rapidly expand its market base throughout the IT industry including the PC and mobile markets, quickly moving to mainstream status, Samsung will continue to deliver next-generation DRAM and green memory solutions ahead of the competition, while contributing to the growth of the global IT market in close cooperation with our major customers.*

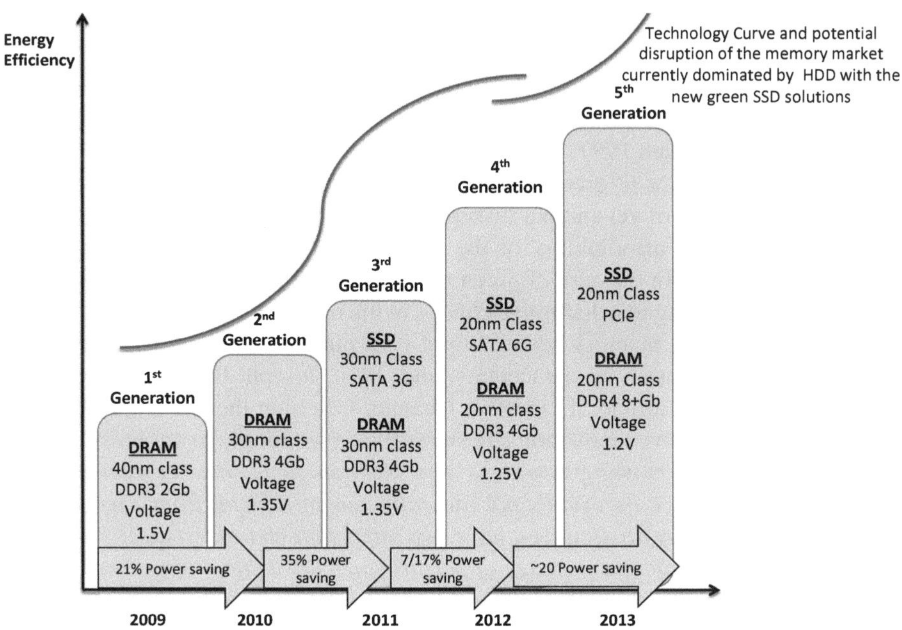

Fig. 4.3 Evolution of green memory products launched by Samsung Electronics (Source: Adapted from Samsung Electronics presentation at CIO Forum 2012)

Section 3: Towards a Disruptive Green Innovation

> *Power consumption in data centers has become a growing concern. In 2010, the world's datacenters consumed 235 billion kilowatt hour, this is equivalent to energy provided by 27 nuclear power plants, and by 2015, 13 more power plant will be required to power the world's data centers.* Excerpts from a talk given by J. H. Lee, VP of Samsung Electronics in 2013

Disruptive Green Innovation

Clayton Christensen coined the term "disruptive innovation" in 1995 to describe innovations that change the nature of the existing market often creating completely new markets. According to him, disruptive innovation describes a process by which an innovation or technology initially takes root in simple applications at the bottom of a market and then relentlessly moves up market, eventually displacing established competitors. In his book, "The Innovator's Dilemma", Christensen takes the example of the data storage industry to showcase the repeating patterns of disruptive innovations that shaped and re-shaped that industry for over 50 years. For example, in 1981, the old 8 inch drives (used in mini computers) were "vastly superior" having six times greater capacity as compared to the new 5.25 inch drives (used in desktop computers). However, at $3000 a unit, the 8-inch drives were not affordable for the new desktop machines. The simple 5.25 inch drive, assembled from technologically inferior "off-the-shelf" components, cost significantly less at $2000 a unit. It can be considered as an "innovation" only in the sense that it was new. However, as this market grew and the drives improved, the companies that manufactured them eventually triumphed while many of the existing manufacturers of eight inch drives fell behind (Christensen 1997).

With the emergence of green memory, it appears that the data storage industry has reached the cusp of yet another disruptive change. In the earlier case of the data storage industry, the affordability of the 5.25-inch drive allowed it to disrupt the 8-inch drive market. In the case of green memory, its efficiency and low cost to the environment is leading to a disruptive change in the industry. In other words, the eco friendliness of green memory seems to put it in place for disrupting the existing memory solutions used in datacenters and PCs. Joseph Bower and Clayton Christensen, in their article "Disruptive Change: Catching the wave", 1995 highlighted how a disruptive innovation can start off as a seemingly unattractive change before invading the existing market. *"The technological changes that damage established companies are usually not radically new or difficult from a technological point of view. They do, however, have two important characteristics: First, they typically present a different package of performance attributes—ones that, at least at the outset, are not valued by existing customers. Second, the performance attributes that existing customers do value improve at such a rapid rate that the new technology can later invade those established markets."* (Bower and Christensen

1995, https://hbr.org/1995/01/disruptive-technologies-catching-the-wave). Thus for a technological change to invade an established market some companies need to adopt and invest in a seemingly unattractive change. With time, other performance parameters catch up to the levels of the existing products in the market allowing the disruption of the existing market.

As a leader in the DRAM and SSD segment of the market, Samsung Electronics' adoption of green memory provides a huge support net allowing green memory to disrupt the existing memory market. Samsung's adoption of green memory technology has its roots in the green management initiative started in 2009. What started as a companywide policy to reduce GHGs and develop eco-products influenced the company's adoption of innovations in the green memory space. Several incremental innovations in fabrication and semiconductor technology led to the rollout of the 5th generation green memory solutions in 2013. Built on 20 nm technology, the fifth generation memory solution includes DDR4 DRAM and PCIe SSD which provide considerable improvements in performance, system installation space and energy consumption. The fifth generation of memory solutions provides a Return on Investment (ROI) that challenges the existing HDD storage solutions. HDDs offer lower cost/GB storage as compared to green memory based SSDs and hence still remain as a popular choice for servers and PCs. However, when parameters such as energy consumption, bandwidth (speed of read write), size and environmental sustainability are taken into consideration, the new generation of green memory solutions emerges as a strong competitor for HDDs.

In the following section, we look at how Samsung built the case for green memory products and why their adoption of green technology may result in a shift of the market towards green memory products.

Data Storage Market

The invention of Dynamic Random-Access Memory (DRAM) in 1966 was one of the core developments that led to the launch of today's computer industry. This invention set the stage for development of increasingly dense and cost-effective memory for computers. The next significant leap in the memory industry was the invention of the Double Data Rate (DDR) RAMS, which as the name suggests allowed for twice the bandwidth of data transfers than what was existing at that time. Over the years we have seen DDR RAMS being superseded by DDR2, DDR3 and currently DDR4 RAMS; each change resulting is more efficient and better performing memory solutions.

From the time when Samsung first entered the DRAM market in 1983, it has been developing next-generation memory solutions. These innovations have since been transformed into energy-efficient, high-performance and large-capacity products that have contributed to the launch of new categories of digital devices. The company's focus towards ceaseless innovation in the memory segment has enabled

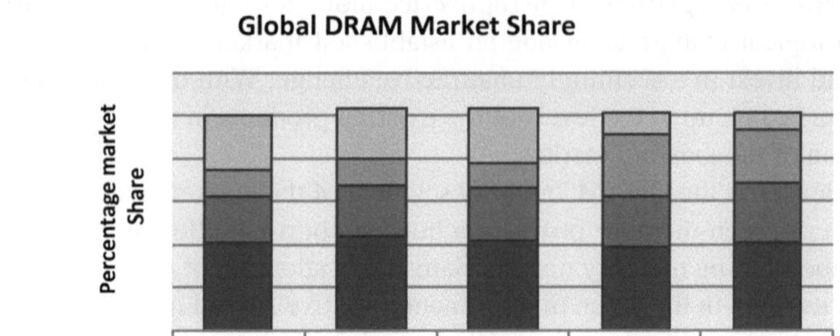

	2010	2011	2012	2013	2014
Others	25.30%	23.70%	25.80%	9.70%	8.60%
Micron Group	12.00%	12.10%	10.50%	28.70%	24.00%
SK Hynix	21.90%	23.30%	25.00%	23.80%	27.70%
Samsung	40.70%	44.30%	42.00%	39.10%	41.40%

Fig. 4.4 Global DRAM revenue and market share (Source: Adapted from DRAMeXchange.com and statista.com)

it achieve and retain the leadership position in the DRAM and SSD markets. As of 2014, Samsung held an impressive 41.4% market share of the DRAM market and 28.5% market share in the SSD market (Fig. 4.4).

While more expensive (per unit storage capacity) than the popular HDD, SSD offer the advantage of having higher bandwidth and lower power consumption. SSDs inherent low power consumption, when coupled with Samsung Electronics' green memory technology makes Samsung's SSDs an eco-friendly alternative to the HDD. Since 2009, Samsung has built a case for green memory solutions like SSD as a replacement for conventional memory solutions based on grounds of environmental sustainability. With the exponential growth of data and data center requirements, the power consumption associated with memory devices has become considerable. The increasing power consumption associated with data storage has made Samsung's case for green memory solutions all the more stronger.

The Case for Green Memory Solutions

The case for green memory solutions becomes apparent when we look at the trend of data growth and datacenter power consumption over the years. An IDC Digital Universe study estimates that the amount of digital data created per year will be 35 zettabytes by 2020. Further, it is predicted that the average amount of storage space acquired by companies will grow by an average of 35–40% per year for external and 33–38% for internal storage. In terms of enterprises' storage spending, the annual average growth rate will be 4.1%, hitting US$42.5 billion in 2017.

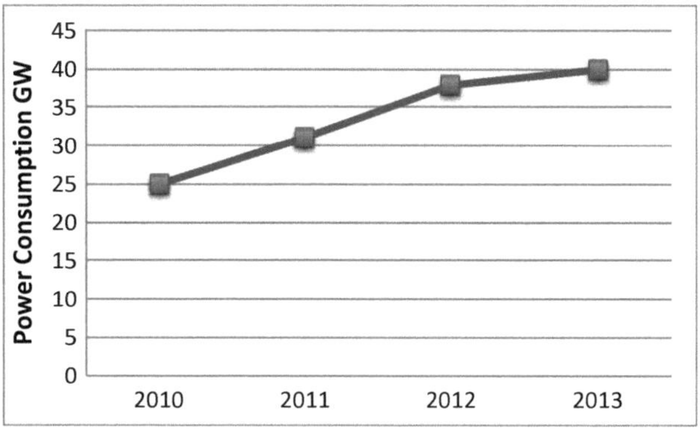

Fig. 4.5 Increase in power consumption in datacenters (Source: DCD Industry Census 2013)

The predicted exponential increase in data and data center requirements has made the power consumption of data centers considerable. Between 2011 and 2012, the global consumption of electricity by data centers went up by 19%. This led to fears that the rising power consumption would continue unabated. (Newman 2014). With increased attention to data center power consumption, companies began to implement energy efficiency measures. These measures soon began showing results. According to DCD Intelligence (DCDi) the rise in power consumption for 2013 was under half of 2012's figures, at just over 7% (Fig. 4.5).

One of the key focus areas of Samsung Electronics' green management initiative was to reduce the power consumption during production and usage of its products. In this light, beginning 2009, Samsung Electronics has focused its attention towards developing eco-products. The company launched the 1st generation of green memory products in 2009 and over the years has adopted innovations that helped reduce power consumption and indirectly GHG emissions during the manufacture and usage of its memory products. In 2013, the fifth generation of green memory solutions emerged that allowed Samsung Electronics to showcase direct ROI benefits of green memory solutions. In addition to performance (faster read, write speeds) and efficiency (lower storage space requirements and low heating) benefits, Samsung Electronics was able to demonstrate calculable environmental benefits from adopting the green memory solutions. By implementing a green memory solution with the same cost as a conventional system, the number of servers could be greatly reduced, the amount of power consumed lessened and the available memory can serve substantially greater number of users concurrently.

With the 5th generation of green memory solutions, a strong case for green memory has emerged. Samsung Electronics' focus towards green management and eco-products has helped the company focus its efforts towards green memory innovations. Ceaseless innovation in the green memory space has resulted in the memory solutions becoming strong contenders to the existing memory solutions (like HDDs) with justifiable ROI. Whether green memory will disrupt the existing memory market depends on certain industry and market parameters. We look at a few of these parameters in the next sub section.

Will Green Memory Disrupt the Data Storage Market ?

Being the market leader in both the DRAM and SSD markets gives Samsung Electronics' the unique opportunity to influence the direction of technology in the memory industry. The company's focus towards eco products and green memory has surely played a role in shifting the direction of innovation in the industry towards green memory products. The 5th generation of green memory solutions by Samsung Electronics for the first time established a ROI that made green memory solutions a viable option for existing data centers. Whether green memory can truly disrupt the existing memory market remains to be seen. This is partly dependent on how strongly green memory solutions can lower costs and continue to produce technological breakthroughs.

Conquering the Cost One of the biggest reasons for the current popularity of 'conventional memory' in comparison to 'green memory' devices is its lower cost per unit storage. According to Forbes (2014), in dollar terms SSDs are over 5 times more expensive than HDDs on average per GB of storage. Traditional storage drives currently cost around 7 to 8 cents per usable gigabyte, whereas even the best-priced SSDs cost over $0.40 per usable gigabyte. As companies like Samsung Electronics continue to invest in green memory technology, the price of the green memory solutions like SSDs are bound to fall. In fact the price gap between HDD and SSDs have narrowed considerably over the years and many estimate that SSDs would slowly begin to match HDD in price.

Samsung Electronics' 5th generation of green memory devices have already begun to show increased long term ROI. According to the company, its 5th generation memory solutions based on DDR4 DRAM and a PCIe SSD can bring dramatic performance improvements, reduce space for the installation of a computer system, and save energy. The green PCIe SSD Solution offers data transmission speeds of 4 gigabytes per second (GB/s), six times faster than the older SATA SSDs. The solution maximizes the ROI by increasing energy efficiency more than 2.6 times. With built-in DDR4 and a PCIe SSD, the high-efficiency green memory solution can improve system performance more than 1.6 times compared to a server system made up of DDR3 and HDD. In particular, the solution can minimize investment costs, reduce power consumption and space for installation by quadrupling the capacity of storage systems through a technology that removes redundant data.

As companies continue to focus on green solutions, innovations that help make products sustainable would be able to achieve better ROI in comparison to conventional solutions. While some analysts predict that green memory solutions like SSDs are unlikely to beat HDD in terms of pure cost/GB in this decade, green memory based SSDs will continue to show better ROI when parameters like performance and power consumption are taken into account.

Conquering the Technology A second key bottleneck inhibiting adoption of green memory solutions is the fact that the green memory solutions are not able to provide all the functionalities that are provided by conventional memory solutions. In specific, green memory solutions generally have lower storage capacities as compared to conventional HDD solutions. According to DCD Focus magazine (2014), most SSD manufacturers focus on SSDs with 256 GB or less. Very few manufacture SSDs with 512 GB, and even fewer manufacture solutions above 1 TB. As a result, while prices for SSDs up to 256 GB continue to decrease, price for SSDs above 1 TB is uncompetitive. This price difference means that while small and mobile devices can increase SSD usage, large storage and enterprise level systems presumably would continue to use HDD or tape storage (DCD Focus 2014). However if we visualize the future scenario where more investment is being put in sustainable products as compared to conventional products, green memory technology is bound to catch up with that of conventional memory.

Samsung Electronics continues to exploit technology breakthroughs that help improve the energy efficiency of its products. For example, Samsung Electronics has begun the mass production of 3D vertical NAND memory (V-NAND), which reduces power consumption and can last ten times longer as compared to existing planar NAND memory. This V- NAND innovation has significantly improved productivity of green memory SSD solutions. With mass production of V-NAND memory solutions, large-capacity (512 gigabyte or 1 terabyte) green memory products are slowly becoming affordable. Thus, although the current green memory technology lags behind conventional memory solutions in areas like storage capacity, the push for green memory innovations meted by companies like Samsung Electronics has helped reduce the technology gap between green and conventional memory solutions.

With companies like Samsung Electronics pushing for green products, we can expect green innovations to become the dominant force in the memory industry. The disruption of the current data storage market with high efficiency green memory solutions are dependent on how quickly these solutions can narrow the cost and technology gaps that exist in comparison with conventional memory solutions. What is encouraging is the fact that the disruption is partially driven by companies that need to ensure environmental sustainability. As long as industry bodies like EICC set norms for sustainability and companies implement management initiatives to adhere and surpass these norms, we can expect such green innovation to emerge and disrupt the conventional market. The data storage industry has for more than 50 years provided a petri dish to study disruptive innovations. The emergence of the green memory provides another beautiful case of a potentially disruptive innovation partially driven by sustainable efforts of companies in that industry.

Challenges Ahead for Samsung Electronics

Samsung Electronics' investment in green memory devices is slowly reaping benefits for the company. The entire story that started with the creation of the EICC in early 2000 is depicted in Fig. 4.6. However good the story may appear, the company is faced with some big challenges. Some of these challenges are highlighted in the sub sections that follow.

Nature of Industry Samsung Electronics has placed a large bet on green memory technology. In 2011, Samsung Electronics increased its stakes in the green memory technology by selling its HDD business to Seagate. This sale allowed the company to concentrate its efforts on green memory technologies. While it appears that the company has chosen the right path, technology breakthroughs can change the nature of the game at any given moment. For example, a breakthrough in the conventional memory technology that allows HDD to operate at very low power can shift the market away from Samsung Electronics' green memory products. In order to counter this challenge, the importance of ceaseless innovation cannot be under stated. It is imperative that the company stays ahead of technology leading the innovations in the memory market. The more effort the company places on ceaseless innovation, the more will it be able to control the direction of innovations in the future memory market.

Competition Compared to the HDD market, which only has three main manufacturers (Seagate, Western Digital, and Toshiba), the list of SSD manufacturers is considerably longer. There are about 40 companies manufacturing and selling SSDs, ranging from computer industry giants like Samsung, Intel and Seagate to much smaller start-ups. As a result, while some of the larger manufacturers (such as

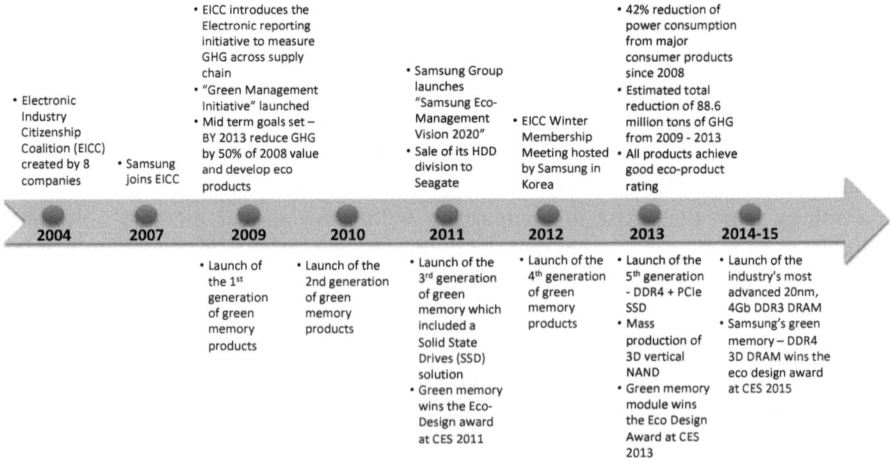

Fig. 4.6 Timeline of important events (relevant to green memory) at Samsung Electronic

Intel and Samsung) can offer some support or integration benefits, the key selling point for most manufacturers remains price, especially in terms of cost/GB. As a result, SSD prices and in general prices of green memory products are being forced down. Falling prices of green memory technology can help it disrupt the conventional memory market allowing for high revenues for the players in the market. On the other hand competition can lead to very low margins and fragmented market share. Again the best course to take to overcome the issue of competition is to innovate ceaselessly. Samsung Electronics, so far, has managed to maintain the market leadership in both DRAM and SSD markets. With its focus on ceaseless innovation and potential success of the green memory initiative it is likely that Samsung will retain its market leadership for the years to come.

With more than 45 years of history, Samsung Electronics' has established itself as a strong contender in the information technology industry. While the challenges discussed above are a source of concern, years of experience and market leadership provides Samsung Electronics with the skills required to address them. The company's rich experience and the unique case of green memory are a source of several insights of interest to both CSR practitioners and managers of technology companies.

Conclusion

Samsung Electronics' commitment to CSR practices has not only had a positive influence on the company but has also shown signs of impacting the entire memory industry. What started as an initiative to improve environmental sustainability of the company has resulted in green innovations and eco-products. In 2009, the company started the green management initiative to contribute to building a prosperous society and preserving the environment. As a part of the green management initiative, focus was placed on innovation and developing eco-products. Accordingly, Samsung Electronics launched its 1st generation of green memory in 2009. Over the years, several innovations in green memory space have led to improvements in energy efficiency and performance of the product. With the 5th generation of green memory solutions, Samsung Electronics has been able to establish a considerable ROI for consumers (especially data centers) to invest in the green memory solutions in lieu of conventional solutions like HDD. With the exponential growth of data due to emergence of virtualization and big data, power consumption of data centers has become an increasing concern. The green memory solutions offer an opportunity to tackle this concern by providing products with high efficiency and low power consumption.

Will green memory disrupt the data storage or memory market? It is too early to predict the full impact that green memory solutions might have on the existing memory market. The memory industry has been volatile ever since PCs came into prominence more than 50 years ago. Innovations in semiconductor fabrication and memory technology tend to rock the industry several times a decade. Whatever may be the future of this industry, the Samsung Green Memory case is a testament of the potential large-scale impact of sustainable practices. As a market leader in the DRAM and SSD memory segment, Samsung Electronics' efforts towards green memory have surely shifted the industry towards considering and adopting green innovations. Samsung may want to go to the extent of saying that with green memory they have created a "disruptive green innovation". Even if this sounds optimistic, we are bound to see increasing cases of "disruptive green innovations" that have the potential to not only change the existing market but also create new markets for environmentally friendly products.

References

Bower, Joseph. 2002. Disruptive change. *Harvard Business Review* 80(05): 95–101.

Bower Joseph L., and Clayton M. Christensen. 1995. Disruptive technologies: Catching the wave. *Harvard Business Review*. https://hbr.org/1995/01/disruptive-technologies-catching-the-wave.

Christensen, Clayton M. 1997. *The innovator's dilemma: When new technologies cause great firms to fail*. Boston, MA: Harvard Business School Press. ISBN 978-0-87584-585-2.

Christensen, Clayton M. Disruptive innovation http://www.claytonchristensen.com/key-concepts/#sthash.jj4mr7aM.dpuf.

DCD Focus. 2014. Off the grid?... questions for the future. Published in Data Center Dynamic Focus, 3(33): Pages 16,17.

Elkington, J. 1998. *Cannibals with forks: The triple bottom line of 21st century business*. Stony Creek: New Society.

Forbes report. 2014. A look at the SSD market: Where do seagate and western digital stand? Published by http://www.forbes.com/sites/greatspeculations/2014/12/04/a-look-at-the-ssd-market-where-do-seagate-and-western-digital-stand-2/.

Fryxell, G.E., and R.S. Dooley. 1997. Saving the commons: A behavioural simulation for experiencing the role of collaboration and trust in devising workable solutions to environmental and other social issues. *Research in Corporate Social Performance and Policy*, Vol. Supplement 2, pp. 149–83.

Grant Thornton International Business Report. 2014. Corporate social responsibility: Beyond financials.

Jennings, M.M., and J. Entine. 1999. Business with a soul: A re-examination of what counts in business ethics. *Journal of Public Law and Policy* 20(1): 1–88.

Koomey, Jonathan. 2011. Growth in data center electricity use 2005 to 2010. Oakland: Analytics Press. August 1. http://www.analyticspress.com/datacenters.html.

Lee, Ki-Hoon, and Ji-Whan Kim. 2009. Current status of CSR in the realm of supply management: The case of the Korean electronics industry. *Supply Chain Management: An International Journal* 14(2): 138–148.

Mallot, M.J. 1998. An interview with Keith Davis. *Research in Corporate Social Performance and Policy* 15: 241–250.

McKinsey Quarterly. 2008. How IT can cut carbon emissions.

Newman, Henry. 2014. SSD vs. HDD Pricing: Seven myths that need correcting. Published in http://www.enterprisestorageforum.com/storage-hardware/ssd-vs.-hdd-pricing-seven-myths-that-need-correcting-3.html.

Wokutch, R.E. 1992. *Worker protection, Japanese style: Occupational safety and health in the auto industry*. Ithaca: ILR Press.

Wokutch, R.E., and M.J. Mallot. 1998. Capturing the creation of a field: Documenting the history of social issues in management. *Research in Corporate Social Performance and Policy* 15: 207–212.

Referenced Public Articles and Report

"Samsung Unveils 5th-Gen Green Memory Solutions". 2013. http://www.businesskorea.co.kr/article/1733/next-gen-green-chip-samsung-unveils-5th-gen-green-memory-solutions#sthash.PjUTHLrT.dpuf.

Product report and factsheet- Samsung's 5th Generation Green Memory Solution; 2013.

Samsung Electronics sustainability report 2010.

Samsung Electronics sustainability report 2011.

Samsung Electronics sustainability report 2012.

Samsung Electronics sustainability report 2013.

Samsung Electronics sustainability report 2014.

Chapter 5
ZTE (CN) Case – Eliminating Digital Chasm

Lam T.M. Eric, Lloyd Alison, and Chen Jianzhou

Introduction

The continuous fast development of information and communication technology (ICT) has brought ZTE a lot of challenges and opportunities. As a leading global provider of integrated communications solutions, ZTE puts a lot of resources on research, development, and innovation to maintain the developing innovative communication technology so as to increase the value for their stakeholders. ZTE designs and implements environmental protection and energy saving solutions and a low-carbon development model. Their services help people to enjoy the high quality and freedom of communication that allow users around the world to enjoy various kinds of communication such as voice, data, multimedia, and wireless broadband and promote the sustainability of the economy, society, and environment. In view of ZTE's development in the past 30 years, sustainable development and corporate social responsibility are the important facets of the organization culture.

As a member of the UN Global Compact's Global e-Sustainability Initiative, in particular, ZTE has focused effort on eliminating the digital chasm in Africa. Since ZTE constructed Ethiopia's national network (Next Generation Network Program) in 2007, the GSM network capacity of Ethiopia increased from 1.2 million lines to 20 million lines in 2013, with the number of mobile users reaching 19 million. With the well-developed telecommunications infrastructure, Ethiopia's annual GDP growth rate exceeded 10%, the fastest growing economy in the past 6 years among the Sub-Saharan Africa Non-Oil and Mineral Economies.

L.T.M. Eric (✉) • L. Alison
Department of Management and Marketing, Faculty of Businessl, The Hong Kong Polytechnic University, Hung Hom, Hong Kong
e-mail: eric.lam@polyu.edu.hk; Alison.lloyd@polyu.edu.hk

C. Jianzhou
ZTE Corporation, Shenzhen, China

G. Lenssen et al. (eds.), *The Role of Corporate Sustainability in Asian Development*, Advances in Business Ethics Research 7,
DOI 10.1007/978-3-319-45160-2_5

Core to their strategy is the emphasis on the need for innovation. ZTE invests more than 10% of its total annual revenue into research and development (R&D) every year. To date, ZTE has established 18 R&D institutions in the world, with nearly 30,000 domestic and foreign employees working on technological development and innovation in the telecommunications industry. ZTE has more people performing R&D work than any other listed companies in China.

ZTE retained its global top-2 position in patent applications in the World Intellectual Property Organization's annual rankings published in 2013, after being the top-ranked company in each of the previous 2 years.

Background of the Company

ZTE was founded as Zhongxing Semiconductor Co., Ltd in Shenzhen, China, in 1985. In 1993, Zhongxing Semiconductor changed its name to Zhongxing New Telecommunications Equipment Co., Ltd The ZTE Corporation of being publicly traded having made an initial public offering (IPO) on the Shenzhen stock exchange in 1997 and another on the Hong Kong stock exchange in December, 2004.

While the company initially profited from the sales of telecom equipment, it used money from its 2004 Hong Kong IPO to further expand R&D, innovation, overseas sales to developed nations, and overseas production. It moved in the direction of the international telecom market in 2006.

- In 2014, ZTE 4G shipments increased 100% and achieved double 4G shipments consecutively. ZTE also maintained the world's fastest-growing company of 4G technology in 2014 as well as in 2013. By December 2014, ZTE has concluded 170 LTE/EPC commercial contracts. With world-leading LTE end-to-end solutions and the good cooperation with global markets, ZTE has entered into nearly 70% countries that have invested in LTE networks.
- Currently, ZTE is the main contributor and participant of the global 5G standard research. At the world's first session of 5G forum, ZTE initially put forward Pre-5G concept which has been a hot topic. In November 2014, ZTE and China Mobile completed the Test of World's First Pre5G Massive MIMO Base Station. ZTE launched the industry leading 4G Qcell indoor deep coverage solution; ZTE firstly accomplished the 4G LTE network stress tests successfully; it also firstly launched the four-mode BBU to achieve integrated networking among different network types; the solutions such as Cloud Radio, UBR base stations have been honored several awards at home and abroad

ZTE claims to devote 10% of its annual revenue on research and development each year, and it could be producing patents and utility licenses at a fast pace. By June 2014, ZTE had filed applications for more than 53,000 patents globally, with more than 17,000 granted so far. ZTE is a global leader with more than 800 essential patents on the standardization 4G LTE.

To respond to global challenges and create a sustainable future, stakeholders need to work together. ZTE is a member of the United Nations Global Compact, Global e-Sustainability Initiative (GeSI), Green Touch, and more than 70 international standards organizations. They cooperated with international organizations, worked together and shared experiences with their partners in sustainable management, and kept improving and promoting their global corporate social responsibility.

ZTE works closely with suppliers, and together with them sharing, disseminating and promoting the importance of corporate social responsibility. In 2013, they provided CSR training for around 200 suppliers, around 900 suppliers' managers and CSR technicians. ZTE requires their suppliers to establish an effective CSR management system and become more socially responsible.

Ideas of 'Tomorrow Never Waits'

This is an era of IOE (Internet of Everything). Brands will be eliminated if the company does not keep up with the current pace and changes. Thus, in order to have quick access to the new era, ZTE has to keep changing and constantly adjust the pace and direction to move towards.

For carriers this means: Creating and achieving a new profit model

ZTE not only provides products for the carriers, but also comprehensive and customized solutions.

For enterprises and governments this means: building a cutting-edge information platform for the smart city

ZTE will dig deep into the values of information. Through fast actions and precise anticipation of future trends, we will present a city's bright future for the public.

For the consumers this means: presenting the bright future for "you"

This is an era of fast changing information. The boundaries of imagination are continuously expanding. Its combination with technology and the future will be transformed into real and touchable apps in users' hands.

1. Create value for customers through continuous innovation, and boost the progress of ICT.
2. Use of advanced communications equipment and network technologies to enhance the overall level of communication in developing countries and regions, and promote local economic development, improving people's living standards.
3. Develop smart solutions and products to let more people enjoy the convenience and services provided by ICT, and promote the communications between people around the world.
4. Serve customers with dedication and be committed to them. Provide customers with competitive products and services, focus on their needs, improve customer

satisfaction, protect consumers' health and safety, and protect the information and privacy of customers and consumers.

5. Set up training centers around the world to provide customers with professional knowledge services such as training, consulting, and assessment. Transfer advanced knowledge and realize the localization of talent and technologies.

Ethiopia Communications Enhancement

Communications technology has greatly influenced people's lives and the community. Also, in this technical development process, ZTE faces a number of challenges, such as, in different countries they have varying communication standards and requirements, communications costs are high so that the general public can't afford it, the Internet network should be more popular, and the issues of culture differences and a digital divide.

These challenges can be reduced by the research & development, technology innovation, system design and applications of communication technology. ZTE has been utilizing its own developed communication technologies and systems to improve the communication in developing countries so as to minimize the digital divide and improve people's living.

For Africa, often the bottleneck for socio-economic development is the lack of infrastructure. To operate in the African region, ZTE devoted much effort to developing telecommunications infrastructure, as it has done in Ethiopia. Ethiopia is an ancient country with more than 3000 years of history. In 2013, Ethiopia had around 91 million citizens, but it lacked resources and its traditional agriculture was not advanced. Finding that development of telecommunications is one of the solutions to push the society to be more efficient. Ethiopia had met with some popular telecommunications companies in the world. However, Ethiopia's economy was weak and could not afford the expensive telecommunications infrastructure. Another problem they faced is due to the geographical position of Ethiopia. Ethiopia is located on a plateau, with high mountains and long distances to reach the households, strong solar radiation and wind. However, prolonged exposure to sunshine, radiation, and wind will make the fiber cables age fast. With these reasons, there are only few companies that are able to meet such challenges for installing fiber cables for Ethiopia's telecommunications network.

However, ZTE was determined to meet the challenges and subsequently entered the telecommunications market of Ethiopia in 1996. With ZTE's advanced communications equipment, installation techniques and network technology, ZTE has not only dramatically improved the overall level of communication in Ethiopia, but also created employment opportunities for the local people and trained local employees. Since November 2006 by signing the "Vendor Financing Agreement of NGN Program", ZTE exclusively built the national network of Ethiopia and helped Ethiopia become one of the countries with the highest level of communications in Africa. Within 7 years with ZTE'S efforts, the capacity of Ethiopia's GSM network rapidly increased from 1.2 million lines to 20 million lines. In 2013, Ethiopia had 19 million mobile phone users, and became one of the countries in Africa that had the best quality of telecommunications.

The development of the telecommunications industry greatly contributes to the economic development of Ethiopia. According to the statistics of the Government of Ethiopia, its Gross Domestic Product (GDP) has maintained double-digit growth for several years recently. In 2011, Ethiopia's GDP growth rate was 11.4%. In the past 6 years, Ethiopia has become one of the fastest growing economies in the sub-Saharan Africa non-oil and mineral economies. African governments and telecom operators have welcomed such investments in the ICT industry to address the dire need in upgrading infrastructure

Beyond the development of the telecommunications infrastructure in Ethiopia, ZTE also contributes significantly to the transfer of advanced technologies know-hows, training of local talents that achieving the technology localization and staff localization:

1. Technology localization: to keep the overall development and updating of communication technologies, other than the technical training of staff, customers also need the training on products and applications. ZTE support Ethiopian Telecommunication Corporation (Named Ethio Telecom now) set up the local University CTIT, which provides local staff and customers with scheduled trainings, and evaluates their work skills to meet with the standards and requirements. ZTE proposed a free scheme for training 1000 engineers for Ethiopia in 3 years.
2. Staff localization: in ZTE headquarter of Ethiopia, around 55% of the staff are local employees, while outside the headquarter, the staff localization ratio is much higher in more than 70%. In addition to staff localization, the management team is also gradually localized as the company stipulated that in every team of five employees there should be at least one Ethiopian employee serving as the team leader. To date, ZTE is the first employer of choice for many Ethiopian university graduates, annually attracting much talent to work in ZTE.

ICT accelerates the dissemination of information and improves work efficiency through low-cost transmission of information. From this angle, ZTE is bridging the digital chasm through closer collaboration with governments to provide a variety of solutions to address the challenges of urbanization, education and healthcare.

"ZTE iCity" Smart City Solution

By 2025, it is projected that 60% of the world's population will reside in cities. Urbanization as a trend will bring about transformation to life and mobility. However urbanization is not without its challenges as problems such as food safety, urban crime, traffic jams, traffic accidents, and lack of educational and medical resources are faced. To alleviate these issues, in September 2012, ZTE launched the "ZTE iCity" smart city solution. For the first time in the industry, ZTE introduced the "4I

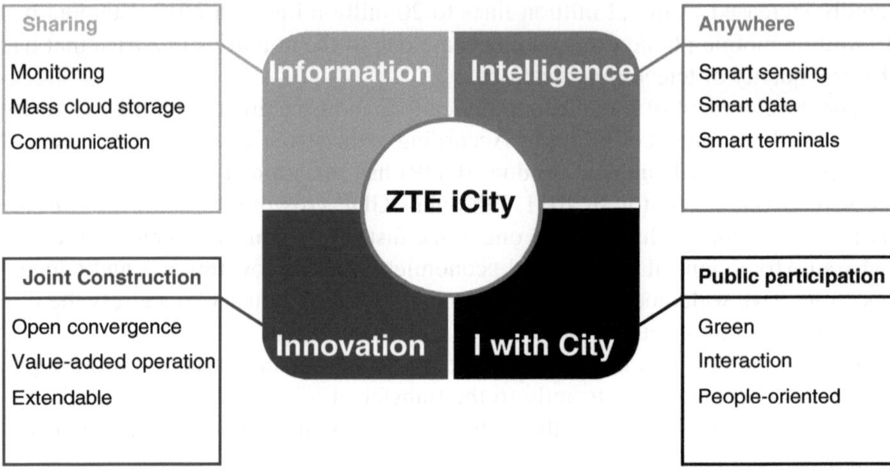

Fig. 5.1 '4I' concept of the ZTE iCity solution

smart city" concept (Information, intelligence, Innovation, intercation), and the cloud computing platform, thus put forward new ideas on modern urban development (see Fig. 5.1). The smart city solution will help improve government management, offer human-oriented services and facilitate people's access to public resources.

"ZTE iCity" smart city solution covers three areas: keeping city stability, promoting city growth, and protecting people's livelihood in cities. They include twelve key applications, including the e-government, safe city, and emergency command, respectively, taking into account one of the urban functions. "ZTE iCity" smart city solution uses the following three areas to help governments manage cities more conveniently, efficiently, and sustainably.

1. Keeping city stability: through the applications of e-government, safe city, digital urban management, emergency command subsystem, city managers can fully grasp the urban economic performance data, public security situation, and police force situation. According to this basic information, city managers can plan and deploy related resources in advance, so as to efficiently respond to extreme weather conditions, public emergencies, and natural disasters, and maintain urban security and stability.
2. Promoting city growth: the construction of the smart city will be the "multiplier" of economic growth, a "converter" of economic development mode, and a "booster" of industrial upgrade. The applications of intelligent industrial park, intelligent transportation, intelligent logistics, and intelligent environmental protection will promote the effective allocation of urban resources, promote the city's economy to develop towards information industry, further developing the new smart city industry chain, and ensure the city's sustainable economic development.

3. Protecting people's livelihood in cities: intelligent healthcare, food safety, intelligent community, and intelligent scenic spot subsystem will enhance the experience of people's lives and allow people to live and work satisfactory.

ZTE smart city solutions have been successfully applied in some 100 cities in the world, including the TD-LTE wireless government network in Beijing, China with independent intellectual property rights, the intelligent industrial park in Suzhou, China, and the safe city application in Marseille, France.

ZTE Smart Education Solutions

Education is an important force for poverty alleviation, improving the overall quality and standard of people's living. For some countries and regions with large population and low levels of Education for All (EFA), unevenly distributed educational resources and great differences between urban and rural areas, the promotion of EFA remains a challenge. Using advanced communications technologies for distance education to improve the cultural quality of citizens and skill levels has an important and far-reaching significance.

Distance education is a modern teaching mode constructed on networks and developed on the basis of traditional teaching. It is an effective high-tech means to promote the modernization of education. With continuous development of distance education, more and more governments, schools and individuals in the world have begun to use distance education system for learning or teaching. Distance education will change the limitations of traditional education mode. Its open education networks will help to instill a sense of lifelong education and creation of a learning society.

In 2013, ZTE launched its iEducation smart education solution. This solution integrates video conferencing, IPTV and video terminals, and disseminate educational information anywhere, anytime through computers, e-book packages and by means of mobile phones, radio, and smart devices, so that users can learn and exchange educational information conveniently. Through various communication methods, the solutions integrate public network and private networks, combine wired and wireless communication technologies, and ensure efficient and stable education network resource sharing. Using an open platform based on Service Oriented Architecture (SOA) and with strong encapsulation capability, the solutions allow the development of new applications and rapid access of third-party applications. Through intelligent analysis of users' learning behavior, the solutions can promote positive interaction of learning. The solutions can also provide knowledge services to build cloud education platform, support professional education, and promote in-school and vocational education.

ZTE smart education solutions provide more learning alternatives for the education sector. With ubiquitous campus networks, an open and flexible software platform, various educational applications and knowledge services, ZTE helps more companies and learners breakthrough the restrictions of traditional education methods, enhance the quality of teaching and training, raise the employment rate, and realize a leap towards intelligent education methods. ZTE smart education solutions have been successfully applied to the educational programs of some countries.

In 2009, the New Guinea Council adopted the 2010–2018 education program, and clearly pointed out that the government will invest heavily in education to upgrade the educational system. To achieve this goal, the New Guinean government decided to construct educational infrastructure and modern digital education platforms in rural areas. The project is based on vocational education, and aims to improve the education level of rural residents. The objectives of this project are to construct a nation-wide data center for digital education systems, with all digital classrooms supporting real-time classroom teaching and non-real-time PC teaching; providing video conferencing functions; and deploying the system in 11 provinces.

With seamless interfacing with traditional education, iEducation solution enhances the teaching experience. It utilizes a unified IT infrastructure, centralized data storage and backup mechanisms. It uses high grade security management, enables users to efficiently share the bandwidth, and is easy to manage and maintain. It deploys classrooms quickly by using a mobile construction program, which is easy to use. It can use flexible transmission schemes to provide rapid urban coverage.

ZTE Smart Healthcare Solutions

The United Nations Millennium Development Goals have three goals that are related with healthcare, including the reduction of child mortality, to improve maternal health, and combating HIV/AIDs,malaria and other diseases. In developing countries, problems such as high cost of healthcare, fewer channels, and low coverage, disrupt peoples' livelihoods. Particularly, a less efficient healthcare system, poor quality of medical services, difficulty to get affordable medical treatment are the main public concerns. The application of information and communication technologies in the medical information systems can greatly reduce the difficulty for patients to get medical treatment, facilitate sharing of information amongst institutions, and improve medical treatment skills in developing countries, and thus help accelerate the achievement of the United Nations Millennium Development Goals.

In developing countries, healthcare can be a difficult business. For instance, in developing countries infrastructure and equipment management of pharmaceutical distribution systems suffer from low efficiency. Developing countries also lack medical logistics management systems, the data exchange and sharing platform for health institutions and the basis and carrier of information integration between different systems. In sum, this translates to inefficient logistics and supply chains, and high expense ratios (in many countries, the logistics costs of pharmaceutical manufacturing industry can exceed 10%). As pharmaceutical logistics have several intermediate links, prices of drugs increase are artificially high. Uneven distribution of medical resources and enormous waste of drugs further lead to lack of medical care, and high incidence of infectious diseases and epidemics in remote areas.

There are additional challenges facing medical information technologies such as the overall low development level and lack of uniform standards. Many hospitals independently construct their own information systems, which do not have unified data structure and format. These systems cannot work with each other, resulting in a waste of resources and money. Meanwhile, the medical information systems have problems such as the protection of users' privacy and information security which prevent sharing of information.

ZTE's pharmaceutical logistics solution integrates information of logistics companies and supply chain information such as cars, drugs, roads, staff and warehouses into a single platform, analyzes the information, scientifically sorts the information, and then reasonably dispatches those information effectively that can reduce the no-load rate, saves warehousing costs, reduces logistics costs, improves logistics efficiency, and reduces carbon emissions. ZTE provides consulting and planning of information systems for pharmaceutical logistics companies before they construct their information systems, and thus brings technological, model, and process innovations for the pharmaceutical logistics industry. ZTE unites the upstream and downstream professional logistics companies in the pharmaceutical supply chain, uses its professional ICT integration capabilities and technical strength to introduce new technologies such as cloud computing, networking, and e-commerce. The solution uses advanced supply chain management information system and equipment to effectively integrate the upstream and downstream resources in the pharmaceutical supply chain, analyze and forecast the demand for drugs, optimize the transportation, warehousing, and distribution management of drug supply. The solution also provides anti-counterfeiting and tracing of drugs to improve the regulation of drugs, so as to achieve the automation, information technology, and benefits of pharmaceutical logistics, and enhance the social medicine level.

Committed to Customers

ZTE conducts its work in products and services by following the core value of "Serving with dedication and being committed to our customers", and protects the rights and benefits of customers. The company keeps customers as the focus of its concerns, executes the TL9000 Quality Management System, and uses 6-Sigma and other methods to improve quality. Therefore, the company has established an overall quality management and improvement system based on customer satisfaction, field operations of products, and internal flows. ZTE constructed the integrated advantages of the company as "Leading Products, Reliable Quality, and Top Services" to continue to provide competitive products and services for customers.

ZTE Global Customer Service Center provides customers with 7×24 technical support. It has 9 product sub-centers, 9 advanced laboratories, 1 technical support team of skilled engineers, perfect technical issue solution banks, an advanced analog laboratory environment, fast and effective control and use of global technical resources, ensuring that ZTE's global customers can enjoy technical support services in a convenient and quick manner.

ZTE is dedicated to constant improvement of its capacity of global customer service, and it has gradually built 8 Regional Customer Service Centers (RCSC) around the world, along with 45 Local Customer Service Centers (LCSC). The company has also established technical support service systems consisting of local, regional, and head offices to provide steady localized field support service teams for the overall implementation of customer support service. Through online support, remote diagnosis, field troubleshooting, and other service modes, ZTE delivers quick responses, high-efficiency and high-quality treatment for service requests from customers, to effectively guarantee the safe and steady operations of customers' online equipment.

ZTE provides global customers with request and complaint acceptance channels, such as hotline telephones, fax numbers, email addresses, physical addresses, websites, and B2B. Furthermore, in order to give customers more convenient services, the company has set up a technical support website, which is an Internet-based window, providing customers with technical support services. The website also offers knowledge base, service center, technical forums, technical documents, and other service warranty functions.

ZTE's ITIL-based model has formed a set of complete customer support service management flow systems and an IT system platform. Currently, it aims at fault management, problem management, technical consulting, service changes, version management, service management, network supervision, and other customer support services, thus providing an overall steady flow and regulation system. In 2013, ZTE optimized the customer service process and innovations in the service level agreement customization, open service process requirements, and B2B docking work orders of customers' systems, and thus greatly improved the customer service capabilities.

Health and Safety of Products

ZTE provides customers with competitive products, services, and solutions while at the same time, is very concerned about the health and safety of users, while focusing on information and privacy protection of customers and consumers. As early as 2005, ZTE passed the ISO27001 "Information Security Management System" certification. In 2010, the company began the ISO15408 "Information technology, security technology, information technology security evaluation criteria" certification. In 2011, ZTE achieved the certification, and became China's first enterprise that achieved the certification. The company's products have passed the U.S. Federal Information Processing Standard FIPS 140–2 (security requirements for cryptographic modules) certification regulated by the National Institute of Standards and Technology of the U.S. In addition, the company set up a Product Security Committee, which is responsible for the company's security affairs.

ZTE is concerned about the ergonomic design of products to ensure that all of its products comply with the applicable health and safety standards. ZTE carefully adheres to the product safety design concept based on user scenarios, and uses the concept throughout all aspects, including product planning, design, development, testing, and manufacturing. User scenarios include the product operational environment, user skills, user habits, and user behavior analysis.

ZTE is an important member of quality and safety standardization organizations, such as ITU-T SG5, IEC TC108, and CCSA. It is actively involved in the activities of product quality and safety, and quality and safety standardization all around the world. ZTE formed its own R&D and production standards based on the product quality and safety requirements of Europe, the United States, and IEC, and improved the design review and testing validation.

In order to ensure a safe and reliable product design, ZTE strictly controlled seven safety risks including risk of electric shock, energy mechanical, radiation, chemical and fire hazards, and overheating. ZTE established a complete product safety laboratory, and operated the laboratory in strict accordance with the international standard ISO17025. The laboratory have been accredited by CNAS, UL, Intertek, CSA, and TUV. ZTE can perform product quality and safety tests, such as CE, FCC, UL, ETL, NEBS, CSA,and PSE, and OFTA. The laboratories provide every convenient test and control methods, an important emphasis for ZTE's products quality and safety.

ZTE strictly evaluates product safety in each stage of production, from product R&D through shipping, in accordance with IEC/EN/UL 60950–1 and GB 4943.1. In 2012, ZTE's products passed various certifications, including China CCC, Europe CE, North America UL, ETL, Canada CSA, Nigeria PC, and CB certifications of other countries. Certified products meet the safety certification requirements of more than 100 countries worldwide. In 2012, ZTE contributed lots of efforts to product safety verification and product safety laboratories, which were 22% more than in 2011. The one-time pass rate of product safety tests was more than 95%, which was 3% higher than in 2011.

Looking Forward

ZTE used a multifaceted technical innovation to develop green products, green technologies, and green solutions to enhance energy efficiency. Thus, ZTE helped their customers and other industries to reduce energy consumption, carbon emissions, and the impact on the environment. ZTE fully implemented the concept of green innovation in the network structure, equipment, boards, and chips. Their green innovation technologies, such as SDR platform, ATCA platform, dynamic energy-saving technology, software energy-saving technology, innovative switching architecture, and highly integrated design, can reduce energy consumption by up to 50%. In electriCity, transportation, agriculture, construction, manufacturing, consumer products, and service industries, the application of our green technologies and solutions can maximize ICT's green utility.

ZTE continues to develop innovative communication technologies, so as to create and enhance value for the company, their employees, customers and partners. ZTE services have helped people from different areas to enjoy equal freedom of communication; narrowed the digital divide and promoted sustainable development of the economy, society and environment.

ZTE continues to use their technology advantages to promote the construction of ICT applications in education, employment, healthcare, social security, public safety, chemical industry, construction, agriculture, and other fields, to facilitate the lives of people and to make things simple. ZTE uses a multifaceted technical innovation approach to develop green products, technologies, and solutions to enhance energy efficiency. Looking ahead, ZTE continues to innovate and bridge the digital chasm by deploying ICT to improve urban livelihood, build a healthier economic and industrial structure, and a stable and harmonious city while promoting sustainable urban development.

References

China-based Huawei and Murali Make Headway in Global Telecom Market DigiTimes.com. March 9, 2007.
Chinese CSR in Africa, http://www.sinoafrica.org/en/node/2275.
Chinese firms join ranks of global leaders in patent applications. http://usa.chinadaily.com.cn/business/2012-11/30/content_15974759.htm. China Daily. Retrieved 15 April 2014.
"Going global" in growth markets — Chinese investments in telecommunications in Africa. http://www.ccs.org.za/wp-content/uploads/2012/04/Telecom_Policy-Briefing_final.pdf. Centre for Chinese Studies, Stellenbosch University. Retrieved on 18 April 2014.
ZTE Annual Report 2013. http://wwwen.zte.com.cn/en/about/investor_relations/corporate_report/annual_report/201404/P020140408599365909862.pdf.
ZTE Corporation CSR Report 2013. http://wwwen.zte.com.cn/en/about/corporate_citizenship/report/.
ZTE Corporation Website. http://wwwen.zte.com.cn/en/.
ZTE Devices – Bring you closer. http://www.ztedevices.com/.
ZTE Ranked World No. 1 in Patent Applications for Second Straight Year. http://www.cnbc.com. CNBC. Retrieved 20 March 2013.
ZTE Smart City White Paper. http://enterprise.zte.com.cn/en/banner_management/banners/201404/P020140419657180221402.pdf. ZTE Corporation. Retrieved 20 April 2014.

Chapter 6
BMW i Story: Revolutionizing Sustainable Mobility in Korea

Hyun Jeong Kim, Jong-Dae Kim, and Sazali Abidin

Introduction

September 29, 2013 marked a significant milestone in the history of automobiles when the first mass-produced luxury electric car went off the production line. The BMW i3 was handed over to the German marathon runner, Jan Fitschen, and used immediately as the lead vehicle at the 2013 Berlin Marathon. On November 15, 2013, the BMW i3 was officially delivered to retail customers in Europe. The Head of the Bavarian State Chancellor's Office and State Minister for Federal Affairs and Special Tasks of Germany, Christine Haderthauer, and Member of the Board of Management of Allianz SE, Oliver Bäte, were among the first owners of a BMW i3. In less than 6 months since then, Charles Rabie, a Professor at Tufts University, who is also a Boston resident, became the first owner of a BMW i3 in the U.S. when he took delivery of the car at a BMW Boston dealership on May 02, 2014.

On April 24, 2014, the BMW i3 was launched at Dongdaemun Design Plaza in Seoul, Korea. At the launch, the Chief Executive Officer (CEO) of BMW Korea, Hyo-Joon Kim said;

> This is not just the launch of a new vehicle. This is the beginning of the future of individual mobility, and a solution to the environmental issues and urban mobility in megacities.

H.J. Kim (✉) • J.-D. Kim
College of Business Administration, INHA University,
100 Inharo, Nam-gu Incheon 22212, Korea, South Korea
e-mail: kimhj@inha.ac.kr

S. Abidin
Lincoln University, Lincoln 7647, New Zealand
e-mail: sazali.abidin@lincoln.ac.nz

G. Lenssen et al. (eds.), *The Role of Corporate Sustainability in Asian Development*, Advances in Business Ethics Research 7,
DOI 10.1007/978-3-319-45160-2_6

The milestone of being the first mass-produced luxury electric car does not come easily to the BMW Group. At the initial stage, the group's current fixation with the internal combustion engine has placed considerable risk on the development of BMW's electric cars. Realizing this threat, the board of directors of BMW Group created a completely separate business unit called the BMW Project i. The technical development team of BMW has been concentrating on the internal combustion engine for almost 50 years and the people are in their comfort zone. Electric cars are new to them and they have little or no experience. Therefore, unlike other car manufacturers, which have adapted existing internal combustion engine to electric cars, the BMW Group created a completely new platform with the BMW Project i to develop a mass produced electric car from scratch (Chater 2014).

The project took 7 years and multi billion dollars of investment before the group finally launched the BMW i3 at the end of 2013. With the huge commitment on time and finances, the return on investment on the BMW Project i is uncertain and it may take many years or even decades before the BMW Project i can recoup its investment costs. The huge research and development expense and the uncertain outcome are the main reasons why other car manufacturers are reluctant to mass produce electric cars. The BMW Group can do so because of the strong commitment of its board of directors towards sustainability. Although the BMW Group is a public-listed company, it is controlled by the Quandt family. The Quandt family aims for the BMW Group to be a sustainable company for many decades to come and is willing to sacrifice huge profits in the short term to ensure the long-term sustainability of the group. This commitment towards sustainability is embedded in the BMW Group (Reithofer 2014a).

The BMW Group is one of only few multinational corporations that have a Sustainability Board, which is composed of the BMW AG Board of Management. The group has a Sustainability Circle that reports to the management board, is composed of nominated representatives from each division and is chaired by the BMW Group representative for Sustainability and Environmental Protection. The individual divisions are responsible for the implementation of strategy, processes and projects to realize the objectives and ensure that sustainability is implemented throughout its value-add chain (Reithofer 2014b).

Having sustainability at the very forefront and defined by the use of renewable resources and innovative means of production, the BMW Group is revolutionizing sustainable mobility. This brings the car industry to an entirely new dimension, where tailpipe emission-free vehicles are a standard way of life, and will be the future of mobility.

With the commitment shown by BMW towards sustainability, this study generates the following research questions:

- What is the motivation for BMW in developing the revolutionizing sustainable mobility in South Korea?;
- What are the factor determinants to the success of sustainable mobility?; and
- How sustainable mobility can improve standard of living and promote good healthy lifestyle of people in South Korea?

Electric Vehicles (EVs)

Electric vehicles (EVs) are currently seen as an important means to make transport more sustainable. Thus far, however, only a few EVs are actually on the road. One study found that the likely early users of EVs will be highly educated middle-aged men. In addition, those who live with their family in rural or suburban regions appear to be more interested in owning an EV (Dütschke et al. 2013).

Greenhouse gas (GHG) emissions from anthropogenic sources are the leading contributors to global climate change. Over the past century, GHG emissions have increased tremendously due to deforestation and the burning of fossil fuels. In the US, approximately 28% of these GHG emissions come from the transportation sector. By replacing conventional gasoline powered vehicles with plug-in hybrid (PHEV) and battery electric (BEV) vehicles, the amount of GHG emissions released can be reduced. To make the transition to alternative fuelled vehicles, consumers must be informed of the economic and environmental consequences of purchasing such a vehicle. A study compared PHEVs and BEVs in terms of both GHGs and cost (Ghosh 2014) found that BEVs produce approximately 4000 kg fewer lifecycle GHG emissions than PHEVs over the life of the vehicles using the US average electricity grid mix. In addition, a lifecycle cost comparison was performed to calculate the payback periods of PHEVs and BEVs when compared with a conventional gasoline vehicle. The analysis showed that the BEV has the lowest lifetime costs due to its increased fuel and maintenance cost savings. With the $7500 Federal tax credit, the payback period of a BEV will only be about 4 months, as opposed to between 2 and 9 years for a PHEV, depending on the specific model purchased.

As the demand for sustainable, low-carbon driving solutions is increasing, the concept of electric vehicles is the next big milestone for the automotive industry. Vehicle manufacture, power grid operators and energy companies are devising a range of approaches to integrate electrical vehicles with the power grid. On the other hand, these approaches need to be compatible, secure and privacy-preserving. Various studies have investigated the security and privacy challenges of electric mobility and focused on the design, implementation, and evaluation of a privacy-enhancing charging solution for electric vehicles. One study suggested that connecting electric vehicles to the energy grid and the Internet gives several advantages to the driver, vehicle manufacturers and grid operators (Hofer 2013).

One study examined the psychological dynamics underlying the charging behavior of EV users (Franke and Krems 2013). Data from 79 EV users in Germany were assessed in a 6-month EV field study. On average, the users charged their EV three times per week, drove 38 km per day, and they typically had a large surplus of energy remaining upon recharging. Based on the first findings regarding the charging style among mobile phone users, the study hypothesized that the user–battery interaction style (UBIS) is a relevant variable for understanding the charging behavior of EV users. That study found that UBIS and comfortable range explain the charge level at which people typically recharged. UBIS is related to the users' confidence in their mental model of range dynamics, the utilization of range, and to the excess energy from renewable sources (Franke and Krems 2013).

The short cruising range due to the limited battery supply of current EVs is one of the main obstacles for a complete transition to E-mobility. Until batteries with a higher energy storage density are developed, it is of utmost importance to deliberately plan the locations of new loading stations for best possible coverage. Ideally, the network of loading stations should allow driving from anywhere to anywhere (and back) without running out of energy. One study showed that minimizing the number of necessary loading stations to achieve this goal is impractical and even worse, the study rules out polynomial-time constant approximation algorithms. Therefore, algorithms with better approximation guarantees need to make use of the special structure of road networks. On the positive side, the study found that by using heuristic algorithms, good solutions can be achieved on real-world problem instances (Storandt and Funke 2013).

China is the fastest growing car market in the world. Both government and industry alike have high hopes for a paradigm shift in mobility towards electric cars. China's industry might be able to catch up or leap-frog once electric cars have hit the roads. Tagscherer and Frietsch described and assessed the current policies in electro mobility in China and placed them in relation to China's scientific and technological capabilities. The results showed that a huge amount of public funding is involved, albeit spread over many programs and provinces. Although the scientific capabilities appear promising, the technological capabilities seem to lag behind (Tagscherer and Frietsch 2014).

The global share of electric cars among new car sales is rising slightly, but remains negligible to date. The improved infrastructure for e-mobility is still a key challenge to the wide-spread acceptance of electric vehicles by customers. The availability of public funding for research and development is falling in all countries except for China, signaling the maturing of the technology in the public eye. Japan is still leading in the production of electric vehicles and battery cells, but South Korea is catching up very fast. Schlick and Bernhart suggested that although the production of electric vehicles has taken place in Europe, competence clusters for the relevant technology (especially batteries) lie in Asia, particularly in Japan and Korea (Schlick and Bernhart 2013).

Sustainability in BMW

One study of the BMW Group illustrated how sustainable leadership can be implemented successfully, even in difficult times. Despite the global financial crisis and the subsequent recession, when other leaders chose to go into survival mode and abandon strategies to achieve long-term goals, BMW stuck closely to its sustainable or "honeybee" principles (Avery and Bergsteiner 2011).

Dr. Norbert Reithofer, Chairman of the Board of Management of BMW Group said in the Board of Management meeting in 2013:

"Sustainability throughout the value-added chain is inseparable from our success and a part of our corporate self-image. That is why we take responsibility. Because of our conviction, but also our self-interest, now and in the future." (Srečanje 2013)

In its most recent Sustainable Value Report (SVR) issued on May 28, 2014, the BMW Group reiterated its commitment to be a key player in shaping the future development of sustainable personal mobility. This is evidenced by the 37% reduction in carbon monoxide emissions of its vehicle fleet to 133 g/km compared to the level in 1995. In addition, the share of renewable electricity in the total electricity supply of the BMW Group has risen to 48%. The report also recommended that the commencement of the mass-produced BMW i3 marked a new dimension of sustainable mobility (Reithofer 2014b).

The BMW Group realizes that the balance between individual mobility and protecting the environment is becoming increasingly important. The group aims to be a car manufacturer that offers the most efficient resource-friendly production and the most visionary solutions for eco-friendly individual mobility. To achieve this, it needs to command leadership in the development of sustainable mobility. Moving forward, the group has set a target to reduce resource consumption during production by 45% per vehicle by 2020, compared to the level in 2006 (Feurer 2012).

Sustainability is a priority throughout the entire BMW i3 value chain. A Member of the Board of Management of BMW AG, Finance, Friedrich Eichiner said:

"The BMW i3 sets a new benchmark for sustainable mobility in all stages of development and production, as well as after sales," (Richter 2014)

BMW has gone to great lengths to insure that the i3's materials, production process, supply chain, and recycling all adhere to sustainable principles. The production of the BMW i3 consumes approximately 50% less energy and 70% less water than other BMW models. The carbon fiber plant in Moses Lake is run entirely on hydroelectric power, and the Leipzig assembly plant is powered by on-site wind turbines. Taking all this into consideration, BMW says the i3's carbon footprint is approximately one third smaller than that of the BMW 118d, which was named World Green Car of the Year in 2008.

The BMW i3 is regarded as one of the most eco-friendly cars because of its use of a portfolio of available sustainable materials, such as the sustainable eucalyptus wood trim the olive leaf-tanned leather. Parts of the instrument panel and door panels are made from natural fibers procured from southern Asia's Kenaf plant. The Chief Sustainability Officer (CSO) of BMW Group, Sean Noonan said:

"Sustainability is part of our corporate strategy – we regard sustainability as a lasting and positive contribution towards the economic success of the company. This is the basis for our ecological and social responsibility." (Chater 2014)

Materials in the car made from renewable resources include that eucalyptus dashboard, Kenaf plant fiber panels and leather seats that are naturally tanned using olive-leaf extract. The BMW i has also been designed so that 95% of its materials are recoverable at the end of its life and the battery packs, which need replacing after about 10 years, can subsequently be reused as temporary storage for solar and wind installations. The company has not yet been able to solve the issue of what to do with the lithium batteries at the very end of their lives, but they have a decade to work with its partner, Samsung of Korea, to find a solution (Chater 2014).

Technological Advancement

The letter "i" in the BMW Project, i stands for innovation, resulting in visionary vehicles and mobility services that carry with them an emphasis on futuristic and modern design. The symbol also equates to a new understanding of "premium" that is strongly defined by sustainability. As such, the BMW Project i is a holistic approach to the mobility of tomorrow, which makes it a truly unique force in the automotive industry. The project provides a much needed breakthrough for the electric vehicle market and helps bring it to scale. The BMW i3 is not the first electric vehicle (EV) in the world but the reason that its launch was regarded as a milestone in the history of automobiles is that it is also the first vehicle of any kind to make such extensive use of carbon fiber reinforced plastic (CFRP) and the first to offer a range-extending gas engine as an option. It is the first production EV from any of the upscale German brands.

In designing the BMW i3, the engineers regard electric cars as a status symbol. They are not meant for long-distance travelers but rather targeted for a mixture of drivers living in cities and commuters with two cars. The vision to mass-produced luxury EV does not come overnight. In fact, BMW's electrification efforts have proceeded in three phases. BMW began field tests of the Mini E, a conversion of the Mini Cooper in 2009. In January 2012, it launched a trial of the ActiveE, which was based on the BMW 1 Series sedan by leasing 1100 vehicles to "electronauts" for a 3-year period who had driven more than 12.5 million miles worldwide. BMW then applied what it learned in both programs to the i project (Russell et al. 2013).

The BMW i3 concept car was first unveiled at the 2011 Frankfurt Motor Show followed by a showcase of its prototype during the 2012 Summer Olympics in London. The production vehicle was officially unveiled simultaneously in New York, London and Beijing on July 29, 2013 while series production for retail customers began on September 18, 2013. As of March 2014, more than 3300 units have been registered in Europe, with Germany, Norway and the Netherlands ranking as the top selling markets. Its potential success to be the preferred electric car is boosted with two World Car of the Year Awards when the BMW i3 was selected as 2014 World Green Car of the Year and also as 2014 World Car Design of the Year. The BMW i3 also won the iF Product Design Gold Award, and in the first UK Car of the Year Awards, the BMW i3 won in two categories, UK Car of the Year 2014 and Best Super-mini of 2014. The BMW i3 was ranked first in Kelley Blue Book Top 10 Best Green Cars for 2014 (Russell et al. 2013).

Designing an EV involves trade-offs among battery capacity, weight and range. The range can be extended by increasing the battery size, but this adds weight, which impedes performance. Similarly, a more powerful motor and power electronics require more energy, which again means heavier batteries or restricted range. A lightweight body, on the other hand, enhances performance, and the weight savings can be "invested" in larger batteries which, in turn, boost the car's range. Unlike the Mini E and ActiveE, the i3 was designed as an EV from the beginning. As such, BMW's engineers could make it extremely light, 2700 lbs. (versus 4000 for the ActiveE). This allowed them to decrease the battery size to 22 kWh (compared to the ActiveE's 32 kWh battery) while offering the same range (approximately 80–100 miles). This not only saves on weight and cost, but also charging time. The i3 has the same onboard charger as the ActiveE, which can charge in 3 h, as opposed to the 4 or 5 h needed for the ActiveE (Russell et al. 2013).

In a previous iteration, the i3 was called the Megacity, which gives a clue to its intended habitat. The car is designed to be compact with a shape that allows it to squeeze into urban parking spots but maximizes the interior space. It also features a 32-foot turning radius, which is feasible because the i3 is a rear-wheel drive. Every part of the BMW i3 has been designed to save energy and weight to maximize range. For example, energy-saving LED lights are used throughout the car. The i3's lithium-ion battery consists of eight modules, each with 12 individual cells, which produces 360 V and 22 kWh of energy. In the event of a fault in any of the modules, it is possible to replace the individual modules. The battery pack sits at the bottom of the chassis, which improves handling and frees up the interior space. The AC coolant also cools the battery, and can be warmed using a heat exchanger, keeping the battery at its optimal operating temperature of approximately 20 ° C. Both the battery and passenger compartment can be preconditioned before a journey begins, while still being connected to the grid (Russell et al. 2013).

The BMW i3 is an electric city car designed for urban use and not for long trips even with the range extender. The BMW Group's experience with test fleets of MINI E and BMW ActiveE cars has shown that a range of 130–160 km is quite sufficient for the vast majority of drivers. The average distance driven daily worldwide is no more than 64 km (40 miles). The BMW i3 achieves a range of 130–160 km (80–100 miles). The vehicle's running costs are comparatively low: Over a period of 3 years, the expected maintenance and running costs for a BMW i3 are about 40% lower than those of the highly-economical 320dA registered in Germany. In other countries, the figures may even be lower, depending on the level of subsidies available for electric vehicles (Sattig 2014).

With a curb weight of 1195 kg it is lighter than most vehicles in the compact segment, yet offers significantly more space for up to four occupants. The car's electric drive-train generates output of 125 kW/170 hp and peak torque of 250 Newton meters, which is immediately available from a standing start. The BMW i3 sprints from 0 to 60 km/h (0–37 mph) in 3.7 s and from 0 to 100 km/h (0–62 mph) in 7.2 s. Its top speed is limited to 150 km/h (93 mph) for efficiency reasons. Charging takes 3 h over a 220-volt line and into a massive 7.4 kW on-board charger, which is one of the largest on the market. BMW claims that charge time can be reduced to 30 min using an optional DC Combo Fast Charging setup (Sattig 2014).

Carbon Fiber Reinforced Plastic (CFRP)

The i3 uses a new architecture called LifeDrive, which is similar in principle to a body-on-frame design. The car consists of two horizontally separate modules, the aluminum module (the chassis), which incorporates the battery and drive system, and the Life module (the passenger compartment), which is made of carbon fiber reinforced plastic (CFRP). With the aggressive weight-saving measures, the i3 motor weighs only 50 kg, whereas the batteries weighs 230 kg. The light weighting imperative extends to components, such as the door trim panels (10% lighter than conventional equivalents), aluminum screws and bolts, and even the windshield wiper blades, which use a honeycomb structure to save weight (Richter 2014).

The CFRP is a high-tech material that forms a significant part of the i's lightweight design strategy. CFRP components are approximately 50% lighter than steel components with comparable properties. According to BMW, the CFRP is the lightest material that can be used in the construction of car bodies without compromising safety. The low weight and high rigidity of CFRP allowed the i3's designers to dispense with B-pillars, so they used opposing "coach" doors to allow easier access to the two rows of seats. The BMW i3 requires no central tunnel, which opens up even more interior space. In addition to being lightweight, the CFRP has other benefits. It saves time and cost on the production line, and it can be repaired more quickly than a metal. Until now, the CFRP body compartments have been manufactured only for special vehicles, such as racers and extravagant individual sports cars. The use of CFRP allows BMW i to set new standards in lightweight construction and completely offsetting the additional weight resulting from high-voltage lithium-ion batteries. At the same time, carbon is a high-tensile material that can be used in a versatile manner for construction purposes and increases the safety of all vehicle occupants (Richter 2014).

Optional Range-Extending Gas Engine

BMW i3 is the EV to offer a range-extending gas engine as an option. The optional range extender is a 650 cc two-cylinder gas engine that is mounted next to the electric motor above the rear axle. The 25 kW (34 hp) gas engine maintains the charge of the lithium-ion battery at a constant level en-route, as soon as it falls below a specified value. This is regarded by many as an interesting innovation. Instead of offering hybrid models with the option to plug in as made by other car manufacturers, BMW offers a pure EV with the option to use gas as backup. Depending on the demand, BMW's production line is flexible enough to build i3s with or without the range extender. The nine-liter fuel tank for the range extender is located in the front section of the car and thus has no effect on the luggage capacity. Apparently, the small motor can produce sufficient energy to maintain the same acceleration and top speed while in range-extended mode. On the other hand, the addition of the extra weight drops the electric range by 10%, and adds another seven tenths to the i3's 0–62 mph time (Russell et al. 2013).

The range extender is powered by the same 647 cc two-cylinder gasoline engine used in the BMW C650 GT motorcycle. The range extender engine only engages when the battery level drops to a pre-specified point, acting purely as a generator to produce electricity to extend the range to approximately 320 km (200 miles) for the European version. The European version has a 9 l gasoline tank while the American version has a smaller 7.2 l tank. BMW clarified that the range extender is designed not for long-distance travel but purely as an emergency backup to keep the electric system going until the next recharging location.

BMW opted to offer a range extender instead of larger batteries for a valid reason. Its study shows that 90% of Americans drive less than 30–35 miles a day. Owing to range anxiety, many users of EV want "backup" power. Having larger batteries will add significantly weight and cost. The range extender and a full gas tank add a mere 450 lbs to weight and $3850 in cost. In addition, the range extender adopts direct current (DC) fast charging, which can deliver an 80% charge in merely 20 min. The range extender increases the car's maximum range in day-to-day driving to approximately 300 kilometers (186 miles). More important than outright range, however, is the fact that with the range extender, drivers can refuel at gas stations rather than having to find an electrical charge station. Gas stations are far more readily available than charge stations and also provide faster refueling (Sattig 2014).

Renewable, Recycled, and Sustainable Materials

Sustainability characterizes the thoughts and actions of the BMW Group, which is evidenced by the BMW Group being the Dow Jones Sustainability Index leader for the eighth year in succession since 2005, making it the most sustainable company in the automotive industry. The attitude of sustainability is enforced from design to production and from the useful life of the vehicle to its disposal. The aim of developing the BMW i cars is not simply to build emission-free cars, but also to use the maximum possible amount of sustainably produced and recycled materials (Reithofer 2014b).

On the exterior, the body of BMW i3 is comprised of two functional units, the aluminum Drive module and the Life module made of high-tech carbon fiber, which uses 25% recycled plastics. The combination of these parts forms an ultra-light, strong design with improved energy efficiency, extended range and greater safety. Remnants from carbon production and the production of carbon components are valuable materials that are either channeled back into the production process or reused in other areas. On the interior, the BMW i3 features door trim panels and a dashboard made from renewable natural fibers, naturally tanned leather, and open-pore eucalyptus wood sourced from 100% forestry certified by the Forest Stewardship Council (FSC). Overall, 25% renewable raw materials and recycled plastics are used in the interior of the BMW i3. The textile upholsteries are made of up to 100% recycled polyester. As a general result of these positive attitudes, 95% of the materials used to produce a BMW i3 can be recycled (Reithofer 2014b).

Consistent sustainability means not only reducing the carbon monoxide emissions produced by the vehicle itself, but also how the vehicle is manufactured and how electricity is generated. As a result of defining sustainability in the early strategic phase of designing a BMW i3, the environmental footprint of the BMW i3 has been certified by the TÜV SÜD Technical Inspectorate. The Declaration of Validity issued by the TÜV SÜD Technical Inspectorate confirms that the methods applied by BMW i are compliant with the latest engineering standards and meet all requirements of the strict ISO standard 14040/14044. These take into account the value chain as a whole, including sourcing, production, usage, and subsequent recycling. To achieve this, the BMW Group is working with sustainable partners, including renewable energy producers, to make electricity from renewable sources. Regarding energy production and consumption, 100% of the power used by the BMW i manufacturing plant in Leipzig is obtained from four Nordex N100/2500 wind turbines, which were installed on site to supply electricity directly for the production process. These wind turbines produce 26 GWh a year with a yearly surplus of up to 2 GWh. The excess power is redirected out to other processes at the Leipzig site. With these wind turbines, the energy consumption and water consumption required to produce a BMW i3 in the Leipzig plant has been reduced by 50% and as much as 70%, respectively. The energy-intensive carbon fiber manufacturing process in Moses Lake, the U.S, can be operated using clean energy from one of the world's largest hydroelectric power plants, the Grand Coulee Dam (Dütschke et al.2013).

E-Mobility

Electro mobility (**E-mobility**) represents the concept of using electric powertrain technologies, in-vehicle information, and communication technologies and connected infrastructure to enable the electric propulsion of vehicles and fleets. Powertrain technologies include full electric vehicles (EV) and plug-in hybrids, as well as hydrogen fuel cell vehicles that convert hydrogen to electricity. E-mobility efforts are motivated by the need to address corporate fuel efficiency and emission requirements, as well as market demands for lower operational costs. E-mobility has now become the key word for debate at the European level (Kraus 2013). The European Union was forced to set ambitious targets to cut emissions in response to the challenges of climate change, energy security and increasing oil prices (Kraus 2013). As a result, car manufacturers have come up with new innovations in areas, such as batteries, which have made electric vehicles a truly viable option for the first time (Kraus 2013).

Comparison on E-Mobility Between Nations

With the emergence of e-mobility, the automotive industry is encountering fundamental changes unlike those seen for decades. Plug-in-hybrids, semi-hybrids, full-electric vehicles, battery technologies, and mobility concepts that have an impact on consumer usage are only a few of the many developments resulting from the emergence of the electric engine. E-mobility primarily involves the use of electric vehicles for different transportation needs and mobility concepts. In a broader sense, the term is associated with a shift to a new network. This network consists of established players from the automotive industry and new players (such as e-mobility and information technology (IT) providers, and battery charging/changing services) that are, in conjunction, shaping the industry with their different products and services. The automotive industry was identified as the most influential driver in the e-mobility network (Dammenhain and Ulmer 2012).

Roland Berger Strategy Consultants, a global strategy consulting firm headquartered in Munich, Germany, and Forschungsgesellschaft Kraftfahrwesen mbH of Aachens, Germany, a firm offering innovative solutions and strategic consulting to automotive industry, formed a collaboration to introduce the E-mobility Index. The index compares the competitive position of the seven leading automotive nations (Germany, France, Italy, the US, Japan, China and South Korea) in e-mobility, focusing on three key indicators: technology, industry and market (Bernhart et al. 2014).

Its latest report on the E-mobility Index for the first quarter of 2014 revealed Japan to be the leader in E-mobility because of its strengths in battery and cell production. On the other hand, Japan's position as the leader in battery and cell production may be compromised in the near future as Tesla Motors of the U.S. announced on May 15, 2014 that it will build a massive electric vehicle battery factory, which it calls a gigafactory make a huge amount of cheaper lithium-ion packs for the EVs of the future. Tesla claims that its gigafactory would be able to reduce the per-kWh cost of a Tesla battery pack by more than 30% and one significant benefit of all these cells coming to market will be cheaper electric vehicles. The automotive lithium-ion batteries to be produced by Tesla are said to have a capacity of 35 GWh and the potential to fundamentally change the structure of the industry. The economies of scale this would create have the potential to transform the entire battery industry, especially when it comes to the business models employed by the manufacturers of large-format pouch and prismatic cells (Bernhart et al. 2014).

Nevertheless, Japan will still benefit from Tesla's gigafactory project as the $5 billion plant will use Panasonic's lithium ion batteries, which are fed with cobalt-rich lithium ion cathodes made in Japan. Currently, Japan is still the leader in terms of new vehicle sales and the technological development of vehicles and functions. This is evidenced as Japanese customers are now paying up to 40% less for a new electric vehicle than their European counterparts. This better value for money leads to a situation in which 80% of electric vehicles in Japan are now being used privately. In addition, Japan is continuously expanding the charging infrastructure for electric vehicles which is regarded as a key driver for E-mobility (Matthies et al. 2014).

In the latest E-mobility Index for the first quarter of 2014, it was reported that Korea has overtaken Germany in this area of technology because of the slight worsening value for money in Germany. Like the U.S., Germany is increasingly targeting the premium segment of the electric vehicle market. France is showing little growth in the range of models on offer, while China and Italy have not shown any technological progress. Japan leads the rankings for industry and followed by the U.S. Both nations remain the most important production locations for E-mobility. Japan profits mainly from its strengths in cell production where by 2016, Japan will account for approximately 60% of worldwide battery cell production. By comparison, Korea's share will be approximately 16%, and Germany will have just 4%. The key location for the production of electric vehicles remains the U.S. where it is predicted that by 2016, more than 460,000 vehicles will be produced there. From 2016, will be roughly neck-and-neck in terms of annual EV production. Between now and 2016, Germany and France will each produce between 250,000 and 260,000 electric vehicles. As for the market for electric vehicles, the U.S. remains the leading market for electric and plug-in vehicles, with almost 96,000 electric vehicles sold in 2013. On a global scale, electric vehicles currently only account for 0.25% of the total vehicle sales (Bernhart et al. 2014).

Value Chain in E-Mobility

On the value chain for electric cars, there are various practices being adopted by electric car manufacturers in different countries. While original equipment manufacturers (OEMs) normally manufacture their internal combustion engines themselves, there are a variety of approaches when it comes to the production of electric motors. German OEMs either produce their electric motors themselves or via joint ventures with their key automotive suppliers, thus giving them access to know-how. Japanese automakers follow a two-pronged strategy, producing some electric motors themselves but also sourcing from external suppliers. By contrast, no French or American automotive manufacturers produce their own electric motors (Dempster 2012).

On the other hand, the production of lithium-ion battery cells remains almost exclusively in the hands of companies that were not originally suppliers to the auto industry, and most of these firms are located in Japan and Korea. This also explains why Japanese automakers enjoy a significant competitive advantage over its European and U.S. counterparts. The Japanese produce the entire range of components themselves or source them very locally, which means they can cover the entire value chain (Hofer 2013).

The core of most E-mobility solutions is the battery. The limitations of the current generation of batteries are visible in three aspects: cost, energy-intensity and range. The price of the battery pack for EV is very high (between €10,000 and

€20,000). In addition, it has low energy-intensity, which means it demands a lot of space for the energy it can provide compared to a modern conventional fuel driven car. The limited range that a standard battery-pack can deliver (between 150 and 300 km) is the most hotly debated problem for EV.

Despite these three obstacles, the advantages of EV are clear to urban consumers. E-mobility produces very low levels of noise and pollution (CO_2, NO_x, particles) and can be powered by renewable energy sources, thus helping the communities of many city centers in developed and developing countries to reach their climate goals. The Chinese Central Government for example has declared that 'electric transport is the way of the future'. With new regulations and incentives that promote sustainability, it greatly enhances the business case for e-mobility on the national level.

Future of E-Mobility

The global market for electric vehicles is showing positive development. With almost 93,000 electric vehicles registered worldwide in 2013, a total of 150,000 units is already forecast in 2014. As for BMW, the carmaker has predicted sales of some 30,000 units of the BMW i3 in 2014. Owing to the high-demand and supply constraints, it currently takes some 6 months for a car buyer to receive their vehicle (Becker 2014).

By the year 2020, it is estimated that one in two new cars will be either partly or fully powered by electricity (Matthies et al. 2014). That may surprise many market observers, who regard the current hype and numerous battery-powered prototypes as a knee-jerk reaction by an automotive industry under stress. E-mobility, however, is no marketing fancy. A study by Bain & Company presented seven reasons that explain why an industry switch to electric cars in their various forms constitutes a fundamental and irreversible change:

(i) In 10 years at the latest the e-car will be a mass-market product;
(ii) E-cars are being launched as today's new lifestyle product;
(iii) The electric car is not just a product variant but it represents a fundamental system change;
(iv) E-cars do not require expensive infrastructure to succeed;
(v) The available e-car technology today is already "good enough";
(vi) Battery costs will be at a mass-production level by 2015; and
(vii) The electrification of the automobiles is compelling and inevitable. (Matthies et al. 2014)

As for Germany, one million electric cars, preferably of German origin are hoped to be registered on Germany's roads by 2020. This is an ambitious goal that the Federal Government has set itself in order to establish Germany as the leading market for electric mobility, and to catch up with the world-wide competition. One million vehicles completely or partly driven by batteries would constitute some 2%

of all cars in Germany today. This aim appears ambitious considering that there are only 2300 pure electric cars and some 37,000 hybrid vehicles registered in Germany today. To achieve this aim, technical, political, and market-related obstacles need to be overcome as Wessing examined the current mood within the entire automotive sector and its associated industries, and suggests that further support is needed for activities, such as research and development, financing, and service provisions (Wessing 2010).

Urban mobility during the next 30 years will be influenced by changes in the demography of cities across the globe, in conjunction with the related cultural shifts that will affect urban residents' transportation choices. In 2010, according to the United Nations Population Division, the world crossed a significant demographic threshold. For the first time in history, the number of people living in cities exceeded the number living in rural areas, and this trend is expected to continue. The U.N. Population Division estimates that between 2010 and 2050, the world's urban population will grow by 80% from 3.5 billion to 6.3 billion. During the same period, the number of people living in rural areas is expected to decline by about 18% from 3.4 billion to 2.8 billion. This growth will pose great challenges for urban mobility, for the networks of transportation facilities, and for services that maintain the flow of people and commerce into, out of and within the world's cities (Moss and O'Neill 2012).

Automobile Industry in Korea

Owing to the improved quality and reliability, and the Korea-US FTA (Free Trade Agreement), the Korean auto industry reached a record in export volume of 3171 thousand units with an export value (including auto parts) of U$71.8 billion in 2012. Despite this, domestic production decreased and domestic sales declined in the 4 years to 2012 as a result of a combination of factors, which include continuous economic uncertainty, withering customers' sentiment and supply shortages caused by the union strike during the 3rd quarter of 2012. On the other hand, sales of imported passenger cars accounted for 135 thousand units or 10.5% of the passenger car market share in 2012, because of the high popularity of European passenger cars with diesel engines, customers' needs for diversity, invigoration of lease financing, and effects of tariff lowering by the Korea-EU and Korea-US FTAs. By the nation of origin, German cars ranked the top spot with 83,956 units or a 63.6% share in the Korean imported market in 2012. This was followed by Japanese cars with 23,855 units (18.1% market share), and U.S. cars with 10,237 units (7.8% market share) ("Korean Automobile Industry Annual Report 2013" 2014).

Production and Sales Review

Domestic automobile production in 2012 declined by 2.0% to 4562 thousand units, which is a downturn of 3 years since 2009 following the Global financial crisis and supply shortages due to the union strike during 3rd quarter of 2012. Vehicle domestic sales in 2012 decreased by 4.3% in 2012 to 1411 thousand units, on account of the decline in economic growth, and higher burden of household debt. Despite the spread of global economic uncertainty due to the continued Euro-zone financial crisis, exports in 2012 increased by 0.6% on a year-on-year basis in 2012 to 3171 thousand units as a result of the improved quality and reliability of domestic vehicles, and rise of competitiveness due to the Korea-EU and Korea-US FTAs. Overseas production of Hyundai-Kia in 2012 leapt 15.8% to 3635 thousand units compared to the former year, which was triggered by the increased production of current plants, such as US, China, India, Czech Republic, and Slovenia, the introduction of new models and local-oriented models, and start-up production in the 3rd plant in China and the new plant in Brazil during the 3rd quarter of 2012 ("Korean Automobile Industry Annual Report 2013" 2014). Figures 6.1 and 6.2 provide a detailed overview of the automotive market of Korea.

On the global front for 2012, Korea maintained 5th place with a 5.4% share of global production following China, US, Japan, and Germany. With China at the top spot in 2012, the nation is ranked the 1st for four consecutive years and in 2012 alone it commanded a 22.8% share in the world's car production at 19,272 thousand units with a 4.6% gain year-on-year. Overall, in Korea, passenger car sales decreased by 1.3% in 2012 to 4.167 million units, buses dropped by 16.5% to 121,000 units and trucks declined by 5.8% to 273,000 units. Hyundai, Kia and Ssangyong saw a positive growth rate, whereas GM Korea and Renault Samsung

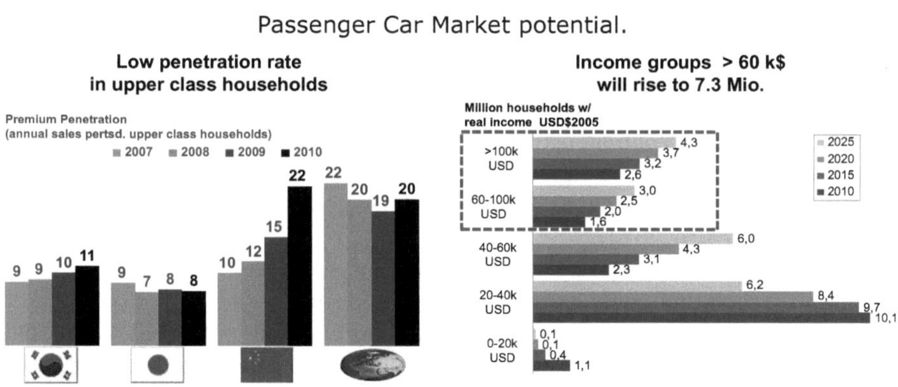

(Source: BMW internal document)

Fig. 6.1 Passenger Car Market Presentation Korea

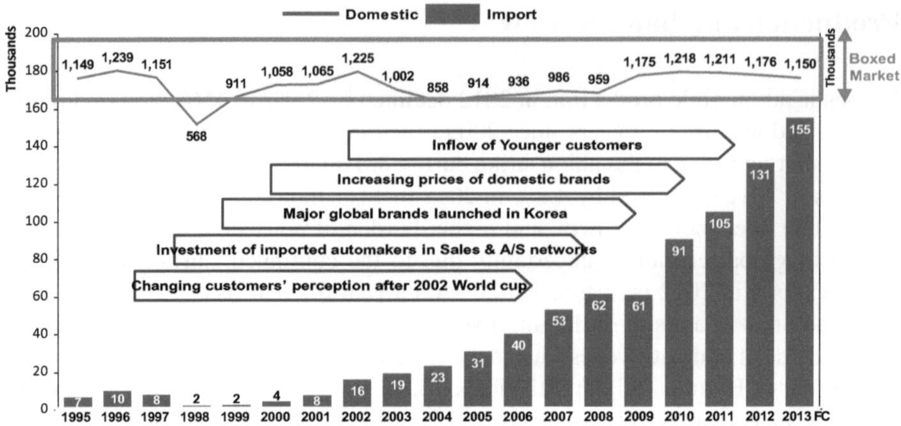

(Source : KAMA / KAIDA, Passenger car only)

Fig. 6.2 Market Overview: Automotive Market in Korea

posted negative growth. Despite the increasing uncertainty over the global economy due to Euro-zone financial crisis, exports in 2012 increased 0.6% and posted 3171 thousand units, showing continuing positive growth, which was triggered by the recovery of auto demand in the emerging markets and the promotion of continuous improved quality and reliability of Korean cars, increasing consumer's preference for fuel efficient small-sized cars amid the high oil price, and rising competitiveness by the effect of the Korea-EU and Korea-US FTAs ("Korean Automobile Industry Annual Report 2013" 2014).

In terms of foreign brand car sales in Korea in 2012, BMW commanded top spot by selling 28,240 units or 21.4% market share in the Korean imported car market segment. With this, BMW ranked 1st for four consecutive years since 2009, when it overtook Mercedes-Benz. The other popular makers are Mercedes-Benz at the 2nd place, followed by VW at 3rd place and Audi at 4th place. The strength of German brands results from the domestic consumers' strong preference, aggressive promotion activities, and price reduction following the Korea-EU FTA. In terms of the model of imported passenger cars, BMW 520d ranked top followed by Toyota Camry, Mercedes-Benz E300, and BMW 320d. Imported passenger car sales are expected to increase continuously due to a combination of various factors, which include the introduction of a variety of new models (diesel engine cars, hybrids, and electric cars), free tariff by Korea-EU/US FTA ("Korean Automobile Industry Annual Report 2013" 2014).

Overseas production, run by Hyundai and Kia, grew 15.8% over the previous year to 3635 thousand units, hitting another new record, owing to an expansion of production in existing plants, such as US, China, Czech Republic, Slovenia, and launching the production of the 3rd plant in China and the new plant in Brazil during the second half of 2012. The 3rd plant in China, which has 40 thousand annual capacity units started production of the new Avante and Santa Fe in July 2012. The new plant in Brazil with a 15 thousand annual capacity began producing the local strategic model HB20 in August 2012. New models, such as the new Santa Fe, Eon and K3 are produced continuously in current overseas plants. ("Korean Automobile Industry Annual Report 2013" 2014).

The automobile industry in Korea provides employment and plays an integral part in the economic development of the nation. The automobile industry in Korea is centered around the major metropolitan areas, such as Incheon, Gyeonggi, Yeongnam district (Ulsan, Busan, Changwon), Honam district (Gunsan, Jeonjoo, Gwangjoo), Choongchung district (Asan, Seosan), and these strategic locations attract many Koreans to work in the automobile industry. In the near future, the Korean automobile industry will be shaped by the increasing demand for hybrids, fuel-cell vehicles, and electric cars. The share of cutting-edge electronic components in automobiles is expected to expand from 32% presently to 40% in 2015 and 50% in 2020 ("Korean Automobile Industry Annual Report 2013" 2014).

Environmental Requirements

The automobile industry in Korea is obliged to adhere to strict environmental requirements, which can be categorized into the following three categories:

Emission Regulations

Automobile makers including importers must obtain approval in accordance with the act indicated by the Ministry of Environment that vehicles are manufactured based on the specific emission regulations and must be able to remain for a guaranteed period of Emission. Petrol/gas-fueled vehicles are subject to the US California's NMOG (Non-Methane Organix Gases) FAS (Fleet Average System) as of 2009.

Fuel-Efficiency

Information on fuel-efficiency should be labelled on the passenger cars, small van and trucks, to enable consumers to select energy-saving vehicles. Fuel-efficiency is indicated as the distance per liter (km/l) and passenger cars are accompanied by its grade.

Recycling

The Korean government tightens the relevant regulations to minimize the environmental impact of the entire process of the automobile industry from production, to use, and scrap. Automobile makers need to solve environmental pollution and the problem of scrap disposal due to the increase in scrap volume, and promote automobile recycling in scrapping. Therefore, automobile makers are prohibited from using lead, mercury, chromium, and cadmium at the manufacturing level, and are required to provide vehicle breakers with disjointing information and increase the recycling rate to more than 95% after 2015 ("Korean Automobile Industry Annual Report 2013" 2014).

BMW Group Korea

The BMW Group Korea (CEO: Hyo-Joon Kim) is the Korean subsidiary wholly owned by the BMW Group, a world-class premium automaker. Since establishing its first import car dealership in Korea in 1995, BMW have been showcasing their concept of premium motoring to the Korean automotive market and provided the best service for luxury cars. As a result, BMW established a firm position as the leader of the premium car market in Korea. The BMW Group Korea now has 8 official BMW dealerships across the country, and more showrooms and service centers than any other import car manufacturer.

To cope with the global trend of rapid diversification, BMW Group Korea is offering the entire lineup of both BMW and MINI family of cars in the Korean market including the BMW 1 Series through 7 Series, the X and M sports versions. BMW is planning to release the new 2 Series and X4, and its line of premium electric vehicles, the i3 and i8, near the end of 2014.

In 2013, BMW Group Korea sold more cars than ever, recording a total of 39,397 cars, including 30 Rolls Royces, which was up 15.5% over the previous year. Specifically, 33,066 BMW's and 6301 MINI's rolled out of BMW showrooms, an increase of up 17.5% and 6.3% respectively. BMW Motorrad also had a stellar year taking first place in the 500 cc or bigger motorcycle segment by selling a total of 1328 motorcycles, up 20% over 2012. In 2013 Korea proved to be the 9th largest vehicle market in the world. Major competition ratio is provided in Fig. 6.3.

In Korea, there are a total of 52 service centers and showrooms for BMWs and 12 for MINIs. In total, the service infrastructure of BMW Group Korea comprises 700 workbays manned by about 1100 well trained service crew members and 35 master craftsmen, which is also more than any other import car manufacturer in Korea. The standardized QMA (Quality Management) dealer system provides the same high level of first-class service anywhere in the country to ensure the best services provided to BMW customers anywhere in the country.

In addition, BMW Group Korea has large facilities in Korea, such as its Vehicle Distribution Center, which helps with the fast delivery of vehicles, and the BMW Regional Parts Distribution Center, which helps minimize the time required to obtain parts needed for repair, all for the sake of customer convenience. After moving the Regional Parts Distribution Center to a larger facility in Incheon in 2006, it immediately became the largest such facility (16,529 m^2) owned and operated by an import car company in Korea. In fact, it serves as a major parts distribution warehouse for not only Korea, but also for a huge part of Asia. The vehicle distribution

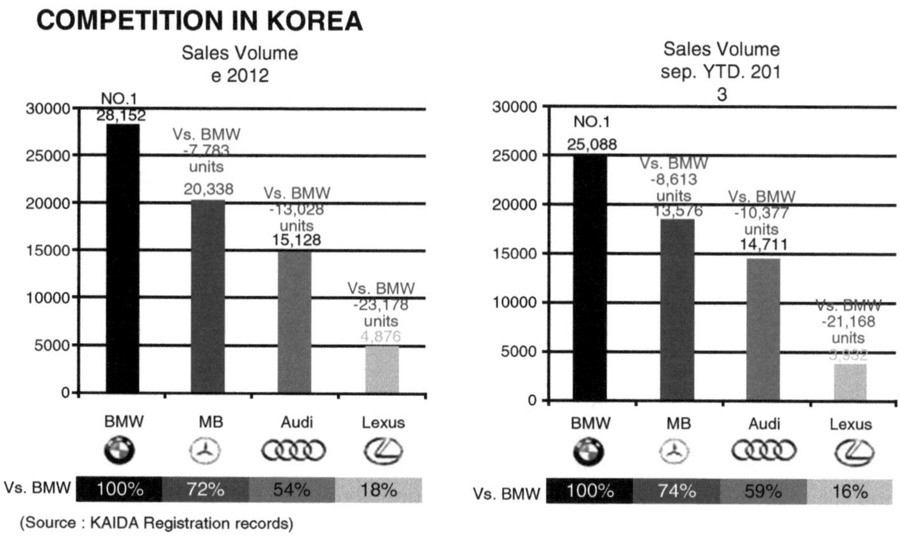

Fig. 6.3 Competition Ratio

center is also the largest such facility owned by an import car company. The center can handle 40 cars a day or around 10,000 cars yearly with staff fully trained to ensure that vehicles are delivered in pristine condition. Of special note, the BMW Training Academy, which was moved to a larger facility in Suwon in May 2007, has become a place for training professional technicians capable of upholding the tradition and reputation of the BMW marque.

BMW Financial Service Korea, established in 2001, develops various financial products, such as leases and loans tailored to the particular needs of each customer. For used cars, BMW Group Korea embarked on a professional used car certification service known as BMW Premium Selection so that customers can have confidence in a previously owned car they drive home in. Buyers can be assured that the vehicles that have passed this program have been through the most thorough and meticulous safety inspection and servicing possible, and are handed detailed documentation for their records.

BMW Group Korea is planning to build the first BMW Group Driving Center in Asia, which will help create a new automotive enthusiast culture and lifestyle. The center will be a multi-purpose cultural space where enthusiasts can revel in automotive culture as well as drive different cars. Various programs will be offered, such as special driving classes, safety classes and structured test drives, so that the customers can more fully understand and enjoy the vehicles that the BMW Group offers. A total of KRW77.0 billion (US$68.95 million with exchange rate US$1.00 = KRW1,116.61 as at March 19, 2015) will be invested in the BMW Group Driving Center that will be constructed on a 240,000 m^2 lot, the equivalent of about 33 football fields. A total of 6 kinds of safety driving education based on the themes of Experience, Joy and Green will be provided. The driving center will consist of tracks for hosting international races, a family-oriented cultural exhibition and experience spaces where customers can experience automobiles from a range of perspectives. In particular, the safety education driving facility scheduled to be opened in July 2014 will not only be the first such facility in the history of the BMW Group, but also the first such attempt in the world. The facility is expected to attract about 200,000 visitors per year to participate in some or all of the activities and programs on offer.

BMW Group Korea's Sustainable Management Efforts

To fulfill its corporate social responsibilities more systematically and effectively, BMW Group Korea established the non-profit BMW Korea Future Fund in July 2011. With the vision of being a "Responsible Leader for the Future," this foundation is engaged in a variety of social contribution activities. The BMW Korea Future Fund is a matching fund based on donations from all stakeholders, including BMW Group Korea, BMW Financial Service Korea, official dealers, and customers.

In 2012, BMW Group Korea raised a total of KRW3.639 billion (US$3.258 million with exchange rate US$1.00 = KRW1,116.61 as at March 19, 2015) through its social contribution activities in Korea. Of this amount, KRW1.975 billion (US$1.768 million with exchange rate US$1.00 = KRW1,116.61 as at March 19, 2015) was raised through the Future Fund, and KRW1.664 billion (US$1.490 million with exchange rate US$1.00 = KRW1,116.61 as at March 19, 2015) raised through vehicle donations. As the BMW Korea Future Fund begins with donations from customers who purchase BMWs and MINIs, it is not an exaggeration to say that the customers are essential to this foundation. Customers' donations are then matched by donations from BMW Group Korea, BMW Financial service Korea and dealers. Thus far in 2013, approximately KRW3.2 billion (US$2.865 million with exchange rate US$1.00 = KRW1,116.61 as at March 19, 2015) (before settlement) has been raised through donations. Excluding other donations, e.g. vehicle donations, the BMW Korea Future Fund has donated approximately KRW2.0 billion (US$1.791 million with exchange rate US$1.00 = KRW1,116.61 as at March 19, 2015) in total so far.

BMW Group Korea has also donated a total of 67 vehicles for research to universities and high schools with automobile-related departments since 2001. In addition, since 2004, BMW Group Korea has been carrying out the BMW technical manpower development program, which is an industry-university cooperation program that provides technical education and opportunities for students to gain employment. at BMW. A popular annual event of the program is the intelligent model car contest, which has been held since 2003 in an effort to foster talented manpower.

BMW Group Korea is developing a new model for shared growth and cooperation with Korean companies and has been continuously providing support for this model at the global level. The head office of BMW currently has 18 primary vendors

Fig. 6.4 Core Values

in Korea including Samsung SDI, Hyundai Mobis, Mando, etc. who is collectively holding contracts amounting to about KRW3.1 trillion (US$2.776 billion with exchange rate US$1.00 = KRW1,116.61 as at March 19, 2015) (based on Lifetime, for 2011–2020). BMW Group Korea is now planning to expand cooperation to include various types of power-packs for electric vehicles.

BMW Group Korea is fully committed to BMW Group's philosophy on sustainability. Indeed, their concerted efforts that have helped the company achieve constant growth have resulted in their being awarded the Korea-EU Social Benefactor Award in 2011. The company emphasizes being a responsible leader for the future by protecting the environment, cultivating global leadership and spreading out the giving culture (Fig. 6.4).

BMW Electric Cars in Korea

BMW officially launched the i3 electric car in Korea on April 25, 2014. Industry experts expect BMW to play leading roles in the supply of EVs in Korea because the German maker considers Korea among its major EV importers. BMW has been expanding such cooperation with the Korean government as the supplier of EV testing vehicles, which will boost the demand for EV significantly.

The BMW i3 has three variants depending on premium options —— the i3 LUX, the i3 SOL and the i3 VIS, which are priced at KRW58.0 million, KRW64 million, and KRW69 million, respectively (US$51,937, US$57,309, and US$61,785 respectively, with exchange rate US$1.00 = KRW1,116.61 as at March

19, 2015). These prices are approximately KRW3.0 million (US$2685 with exchange rate US$1.00 = KRW1,116.61 as at March 19, 2015) cheaper than the model with the same options in Germany. The entry i3 LUX will be available on the Korean market in the second half of 2014 along with another EV model from BMW, the i8. Although consumers will receive subsidies approximately US$20,000 from the government for the environmentally friendly vehicle, the i3 will still be the most expensive EV in Korea. Other electric cars already on the market are Kia Motors' Soul EV and Ray EV, Renault Samsung Motors' SM3 ZE, GM Korea's Spark EV and Nissan's Leaf.

BMW Korea makes a sales target forecast of 250 units of BMW i3 by the end of 2014 and hope to bring in 500 i3s in 2015. On the other hand, as the BMW i3 gains popularity since its launch, securing volume for the Korean market may be difficult. BMW Korea reported that it has received over 130 pre-orders even before the official launch and despite its high price tag, most orders were made by individual customers instead of corporations and public organizations. Generally, customers have to pay approximately $7000 more to buy an i3 than other EVs in the Korean market. Despite this, the company believes that people are willing to pay a premium for an EV that offers luxury driving experiences that are different from other automakers' products.

The automaker believes that the i3 will appeal to a new target group who are environmentally conscious and also seek the same kind of agile driving that people expect from a conventional BMW. For the BMW i3, the carmaker offers a warranty of 100,000 kilometers or 5 years on i3 parts and guarantees 70% battery life for up to 8 years or 100,000 kilometers.

Along with the launch, BMW Korea said it will continue to set up charging stations for EVs nationwide to help promote the commercialization of green cars in the country. In December 2013, the company donated 30 EV charging stations on Jeju Island and in March 2014, it signed a partnership with POSCO ICT, a leading firm specializing in EV charging stations to construct charging infrastructure. The partnership will set up about 300 charging stations nationwide, including 60 chains of the E-Mart discount store so that EV drivers can have easy access to charging stations.

By donating 30 units of electric car charging stations to Jeju Island, BMW became the first automaker to support Korea as a nation in that way. Jeju Island has become the center of electric vehicles in Korea and is most active in the commercialization of electric vehicles (EVs) compared to the other provinces in Korea. In 2013, the Ministry of Environment introduced 1150 electric cars, including 500 allocated solely to the island, and provided KRW15.0 million (US$13,424 with exchange rate US$1.00 = KRW1,116.61 as at March 19, 2015) in support to those who purchased an electric vehicle. The island was designated as a leading electric-car city by the Ministry of Environment in 2014. The provincial government plans to distribute a total of 500 electric vehicles to the private sector in 2014 and hopes to turn the island into a carbon-free territory by changing the island's 370,000 cars into electric cars by 2030.

Jeju Island is bustling with brand-new electric vehicles (EVs) and all the related technologies that will lead the future of green transportation. The island has a circumference of 180 kilometers, making it suitable for electric cars. Thus far, 497 chargers are installed across the island, averaging one charger every 3.7 kilometers. The Korean government has great interest in the environment and in environmental-friendly policies. In May 2014, the support from Korean government was evidenced by the hosting of the first International Electric Vehicle EXPO in Jeju Island. In the exposition, almost 40 companies in the electric vehicle industry and about 200 experts in the sector gathered on the island to show support for the growing interest in the environment, and to discuss ways to strengthening regulations for the protection of natural world. Furthermore, the Korean government plans to offer incentives for 1000 electric vehicles purchased in 2014 and BMW aims to claim 25% of the government's support in the EV market.

BMW is actively collaborating with local Korean companies in producing parts and local contents for its cars. As mentioned before, there are 18 domestic companies, such as Samsung SDI and Hyundai Mobis, which supply parts to BMW. In December 2013, BMW Korea signed a deal with the IT system provider Posco ICT and the country's largest discount chain, E-Mart, to establish EV chargers. The company plans to set up more than 120 chargers this year at 60 E-Mart branches nationwide. For customers who use a Wallbox, a charger that can be installed at home, BMW Korea said a full charge will only cost KRW1,330 (US$1.19 with exchange rate US$1.00 = KRW1,116.61 as at March 19, 2015) worth of electricity if the Korea Electric Power Corporations' special EV fare is factored in. If an owner drives 1500 km a month in the i3, it will cost only KRW32,500 (US$29.10 with exchange rate US$1.00 = KRW1,116.61 as at March 19, 2015). The Wallbox is sold separately for KRW3.0 million (US$2684 with exchange rate US$1.00 = KRW1,116.61 as at March 19, 2015). On the other hand, a home-use emergency charger, which uses 220 volt power, is offered at no cost.

On the global front, BMW's Chief Executive Officer, Dr. Norbert Reithofer reiterated its aim to sell two million or more cars in 2014, of which Korean sales are expected to make up some 40,000 units. In 2013, the carmaker delivered a record 1.96 million cars worldwide, hitting the No. 1 spot in the premium segment. For the first time, its car sales exceeded more than half a million vehicles in Asia in 1 year at almost 580,000 units. Being the best-selling foreign car brand in Korea, BMW sold 39,558 units of its Mini, Rolls-Royce and BMW cars in 2013, accounting for approximately 3% of the country's total car sales. In particular, with its flagship 7 Series limousine, Korea is BMW's fourth-largest market. Dr. Norbert Reithofer also pledged that the carmaker will strive to grow further by enhancing expertise in future technologies, such as electric mobility and lightweight construction. To achieve this, the company has invested approximately 1.5 billion euro ($2 billion) in vocational training and professional development programs. He further added:

> "In 2016, BMW will mark its 100th anniversary, a historic milestone. We will not be looking in the rear-view mirror but focusing on the road ahead. We will be starting from pole position in creating and defining future mobility." (Reithofer 2014a)

BMW Korea Future Fund

One of its programs that best represents BMW Group Korea's concern for sustainability is the establishment of the BMW Korea Future Fund. Launched in July 2011, the BMW Korea Future Fund is a non-profit organization established under BMW Group Korea's grand scheme to fulfil its social responsibility in a more systematic and effective manner. Grounded in its vision to be a "Responsible Leader for Future," the organization is involved in a wide array of social contribution activities that are funded jointly by BMW Group Korea, its authorized dealerships and BMW customers.

BMW Korea Future Fund's 'Junior Campus' is an onsite children's science and creativity education program for elementary students. In 2013 alone, the program staff visited 110 elementary schools, branch schools and child welfare facilities around the country to provide science and creativity education programs for about 8000 children. For 3 years since its establishment, the BMW Korea Future Fund has also been supporting the 'Hope Sharing School' program that provides children from low-income families with lunches and various activities during the school break. Since July 2013, the Future Fund has also been carrying out the 'Young Engineer Dream Project' that provides mentoring services for youths studying automobile engineering in technical high schools and Meister high schools. The Future Fund also promotes global entrepreneurship among students and provides them with the opportunities to launch venture firms by sponsoring the 'From Idea to Start-up,' a national start-up program for college students.

Junior Campus on the Road Across the Nation for Science & Creativity Education

One of BMW's major programs is the Junior Campus, which is a program whereby specially selected teachers travel the country visiting schools to educate elementary school students in science and creativity. The program is a type of localized version of the children's program that BMW operates in its hometown of Munich named BMW Welt (BMW World in English) with the addition of the basic theories taught through the elementary school curriculum in 2013, which are science, technology, engineering, art, and mathematics (STEAM).

The BMW Korea Future Fund has custom built a vehicle solely for the purpose of visiting small branch schools and children's welfare facilities in rural areas and remote places. In 2013, the vehicle visited 110 elementary schools, branch schools and children's welfare facilities to provide science and creativity education programs to almost 8000 students. This is not just a normal vehicle you see every day. This vehicle is an 11.5-ton truck modified into a science lab so that children can learn the scientific technology hidden inside cars through experiments and experiential apparatuses in a fun and amusing way. Through workshop activities, they can

Fig. 6.5 Mobile Junior Campus

make green cars on their own so they are naturally exposed to the idea of sustainable energy. BMW Korea Future Fund aims to help these bright minds strengthen their interest and attention to sustainable scientific technology, cultivating a sense to be a responsible leader in the future. As shown in Fig. 6.5, with the help of the mobile junior campus, children can learn 'sustainable' science and technology through automotive engineering principles. In particular, it can reach various underprivileged children via the 'mobile' format.

When you board this Junior Campus vehicle, a video is shown, in which the Junior Campus mascots, Mirae (future in English) and Haneuri (sky in English), welcome the visitors and then introduced the basic automotive architecture and engineering principles. Following the video is a quiz session about essential car safety, how energy is produced in the engine then transferred to the drive wheels, how velocity changes based on the gear ratios, principles of suspension dynamics, and the basics of aerodynamics. Last but not least, the course also includes a unit on the importance of eco-friendly sustainable energy.

Talent Share and Mentoring Project for Adolescents Aspiring to Become Engineers

This year, the BMW Korea Future Fund is launching a new, promising program entitled the Young Engineers Dream Project. This project is about sharing the knowledge of engineering that BMW Group Korea and its authorized dealers have with enterprising young minds. The candidates will be those currently enrolled in the department of automotive engineering at industrial high schools and Meister schools who come from low-income families and be given mentoring on their field of expertise and future directions.

No fewer than 12 professional technicians from BMW Group Korea and dealerships will be mentors for a total of 24 high school students. Following the launch ceremony held in November, they will be sharing their onsite experience and accumulated knowledge by meeting with each mentee once a month until October this year. The mentors will focus mostly on helping young minds reach an accurate understanding about automotive maintenance and receive advice on how to strengthen their competencies as professional technicians.

The 24 high school students selected for the program will have a chance to visit the BMW Service Center, where their mentors work and in October, before this program comes to an end, they will visit the BMW plant in Shenyang, China as well as authorized suppliers in the vicinity. Moreover, the program has also scheduled a meeting with the BMW meisters from Germany so that they can gain an even clearer picture of their dreams and aspirations.

From Idea to Start-up: Support Undergraduates' Brilliant Ideas

The BMW Korea Future Fund is also actively engaged in activities geared specifically for undergraduates to help them build a sense of responsible leadership. To achieve this, the foundation sponsors the "From Idea to Start-up" competition to foster entrepreneurship among undergraduates and even the newly graduated nationwide. Supervised by the Ministry of Knowledge Economy and jointly hosted by the Korea Institute for the Advancement of Technology, "From Idea to Start-up" is an open competition for start-ups who are developing promising technologies and innovative business models centered on sustainability. This will be a great way for the participants to develop their entrepreneurship while bolstering their chances for success.

One of the judging process includes the 'Startup Camp' where participants can meet face to face with CEOs of various industries and venture business operators to listen to their advice on technologies and business plans. They also receive advice from industry experts in the fields of finance, taxation and patents, which are essential for any startup business. Therefore, this program is one of the greatest chances for any young mind aspiring to launch a business just by participating.

The winner will be awarded practical start-up incubation support, which includes startup funds, office space and an opportunity to visit global firms. Last year's winners looked around the world's leading companies and research centers in Germany, such as BMW, SAP, METAIO, Fraunhofer, EWS and Wildpoldsried, to gain the first-hand experience needed for global entrepreneurship and learn the business onsite. In addition, this year's winners will be able to visit select venture firms in San Francisco and Silicon Valley to meet with the start-up professionals so that they can further refine their business ideas.

Hope & Share School: A School Where Everyone's Happy

BMW Group Korea held its BMW Korea Future Fund Charity Gala in November, 2011 to raise approximately KRW180.0 million (US$161,037 with exchange rate US$1.00 = KRW1,116.61 as at March 19, 2015) through a charitable auction. These funds were then given to the 'Hope and Share School' to be used entirely for a special winter vacation class for children from low-income families, which included all lunch meals, health check-ups and enjoyable extracurricular activities. The program was a huge success as it reached out to 5600 students from 248 elementary schools throughout Korea. In addition, BMW Korea supported about 100 prospective middle school students with money for school uniforms as a way to demonstrate its good will for future minds.

Green Future Camp –Cultivation of Environmentally-Conscious Future Talents

In June 2012, BMW Group Korea held its 'Green Future Camp,' an education camp on the environment for families with children attending elementary schools during the summer vacation season. This 3-day camp attracted almost 250 families whose children learned how precious our environment is while also gaining a sense of responsibility as global citizens and experiencing eco-friendly technology and lifestyle through diverse fun activities. The parents were not neglected either; they attended the parental guide session to raise their children as responsible leaders and participated in environmental volunteer activities in which every family member played together. This was such a meaningful event where everyone involved learned a global sense of citizenship and the responsible leadership demanded for a sustainable future. In addition, the activities led by the Future Fund Foundation, BMW Group Korea, is unfolding a variety of social contribution activities on its own terms.

BMW Korea Donates Vehicles for Research Purposes

Since 2001, BMW Group Korea has donated 67 cars for research purposes. For this year, as a part of its Apprentice Program, the company has donated a mixed fleet of vehicles including BMW 1 Series, 3 Series, X models, Z4s, and MINIs worth approximately KRW1.0 billion (US$894,621 with exchange rate US$1.00 = KRW1,116.61 as at March 19, 2015) to 18 schools and institutions including colleges, high schools and the Incheon HR Development Center, where the automotive department is installed.

BMW Group Korea has donated a total of 67 cars for research purposes since 2001. The Apprentice Program that BMW has operated since 2004 is an industry-academia cooperative that provides technical training to undergraduates in Korea, who then stand a good chance of being offered employment at BMW. BMW Korea will also continue sponsoring the intelligent model car competition as it has done since 2003 as part of its talent cultivation program.

'E-Mobility: Paradigm Shift and Development Plan' – Joint Conference on E-Mobility with the Ministry of Environment

In 2013, BMW Group Korea and the Ministry of Environment held a joint conference under the theme of "E-Mobility: Paradigm Shift and Development Plan" with the relevant companies, government authorities and academia to boost consumers' awareness of electric cars and encourage their popularity as countries around the world continue to tighten regulations on greenhouse gas emissions. During the interview, Mr. Cho, who is in charge of Corporate and Government Affairs mentioned:

> "It was a meaningful occasion through which the current status of EV technologies and participating industries was discussed, as well as the current trends in consumer markets and the government's policies for the industry."

Because the demands for sustainability have brought about the need to introduce new sources to power automobiles, the necessity for additional social overhead capital (SOC) has been thrust upon BMW Group Korea. As new type of infrastructure needs to be implemented, for example, how to best build-out the power infrastructure required to spur to the growth of the electric car industry, the meeting stressed the importance of a quick charging system and discussed examples of how advanced economies are addressing these emerging issues.

Implications on Sustainability in the Context of Asia

As the world is changing due to climate change, depletion of natural resources, growing urbanization, changing levies on cars, changing carbon dioxide and emissions regulations, changing values in economics, culture and consumer expectations, the BMW Group is changing to better position itself in the wake of these changes. The Group is taking a holistic approach in the implementation of sustainability throughout the value chain and views this as a key driver of its business and an investment in the future. This is achieved by gently and efficiently using resources in the company and acting responsibly towards employees and society. This positive approach towards sustainability is presented clearly and forms an integral part of the

Group's corporate strategies, which can be divided into the following 3 components:

- ecological and social sustainability throughout the value chain;
- comprehensive product responsibility; and
- a clear commitment to conserving resources.

Adopting a holistic approach in implementing sustainability means that the BMW Group also recognizes that its 12,000 suppliers in 70 countries have a significant impact on its sustainability performance and sustainable development of society. Therefore, the BMW Group regards it as important for its partners to comply with the same environmental and social standards through which they measure themselves. This includes respect for internationally recognized human rights, labor and social standards in each of the countries. The standards for sustainability being adopted for BMW operations in developed countries, developing countries, European countries, or Asian countries are the same.

In its efforts to ensure that sustainability is implemented in a holistic perspective, the BMW Group focuses on two main areas; minimize risks and strengthen cooperation with suppliers. The BMW Group continuously develops its sustainability risk management by identifying and analyzing the risk management potential sustainability risks along the supply chain. The measures implemented include a proprietary sustainability risk filter, self-report questionnaire, and sustainability audits. In addition, it also explores the opportunities and potential to strengthen cooperation with its suppliers by providing training to its suppliers and implementing projects to improve resource efficiency in the supply chain.

The success of the BMW Group has always been based on long-term thinking and responsible action. The company is convinced that sustainable development is a key factor for its long-term economic success, and thus the philosophy of sustainability is firmly anchored in the company's strategy and culture. Therefore, the company has established ecological and social sustainability throughout the value chain, comprehensive product responsibility and a clear commitment to conserving resources as an integral part of its strategy. These philosophy and business ethics are enforced not only in its headquarters in Germany but globally to all of its 29 production and assembly facilities in 14 countries and sales network in more than 140 countries. As of 31 December 2013, the BMW Group had a workforce of 110,351 employees. The BMW Group regards its employees as the foundation for its success. Against the backdrop of the world of different challenges in the various markets and countries that the Group operates, its employees earn above average wages according to world standards, on top of social benefits, individual models for retirement or health, and accident insurance. The company promotes cultural diversity in business and offers safe and attractive jobs as well as comprehensive development and training opportunities.

In upholding the Efficient Dynamics philosophy, sustainability has always been the driving force behind how the BMW Group operates and the company constantly seeks intelligent solutions to reduce fuel consumption and carbon monoxide emissions without compromising on sheer driving pleasure. The company firmly believes that electric mobility is the way forward to achieving greater sustainability. The introduction of BMW i as the mobility solution of the future goes beyond producing electric vehicles only but entails innovative inter-modal transportation concepts, such as car-sharing and intelligent applications designed to make individual mobility more efficient.

The BMW Group recognizes the growing importance of the Asian region as sales surged by 15.9% year-on-year to 25,300 vehicles in 2013, which is more than double the 6.4% growth posted worldwide. This is strengthened further by the proposed ASEAN (Association of Southeast Asian Nations) integration scheduled to commence in 2015, which will shape the region as an economic community that is expected to converge and become a single market to promote regional harmonization of standards, environment policies and tax structures. The ASEAN integration is bound to attract more investments from companies all over the world, paving the way for a growing automotive market and a production hub for both car manufacturers and suppliers. On the other hand, the Asian region is not at the same stage of adopting sustainability as western countries. Sustainable mobility for the Asian region initially focuses on hybrid vehicles because electric vehicles are only viable in very few urban areas of large cities in Asia. Electric vehicles require appropriate infrastructure and government support in the form of tax incentives and duties, which encourage the development of sustainable mobility. This is the main reason why the BMW i3 is currently being sold in only 30 countries all over the world including 4 countries in Asia (Singapore, South Korea, China and Japan).

Conclusion

Transforming an automotive industry into a more sustainable-oriented automotive provider has increasingly become a critical issue for firms, governments and customers. As suggested in the above main context, one in two new cars will be either partly or fully powered by electricity by the year 2020 (Matthies et al. 2014). The issues related to battery-powered prototypes, e-mobility, electric cars in their various forms, electric mobility, and green transportation all constitute a fundamental importance to change.

This study documented the experience of BMW's efforts in revolutionizing sustainable mobility in Korea through the BMW i story. The underlying complexity, the scale of the project, and the expectations of the BMW's experience to meet various stakeholders highlight the importance of corporate sustainability, particularly in the context of Asia.

A growing number of organizations have recognized the importance of corporate sustainability in an environment where turbulence and unpredictability are increasingly becoming the norm, organizations without the required capability, in this case corporate sustainability, are likely to experience difficulties in supporting and leveraging their assets. In this regard, a detailed elaboration of BMW is highly relevant to firms in the automotive industry. Overall, this case has enabled us to learn the critical issues in corporate sustainability in the automotive industry, as well as the technical and social difficulties associated with green transportation. In addition, this study highlights the need to consider what is required to achieve the paradigm shift required by various stakeholders, such as government, automotive industry and customers.

References

Avery, G.C., and H. Bergsteiner. 2011. How BMW successfully practices sustainable leadership principles. *Strategy & Leadership* 39(6): 8. doi:10.1108/10878571111176583.

Becker, T. 2014. Making emobility work: The way to sustainable mobility, 44. BMW Group.

Bernhart, W., T. Schlick, I. Olschewski, M. Thoennes, and J. Garrelfs. 2014. E-mobility index for Q1/2014 (pp. 11): Roland Berger Stragey Consultants and Forschungsgesellschaft Kraftfahrwesen mbH.

Chater, S. 2014. BMW corporate news, 38. BMW Group.

Dammenhain, K.-O., and T. Ulmer. 2012. Managing the change to e-mobility, 4. Capgemini.

Dempster, P. 2012. A future vision of sustainable mobility, 21. BMW Group.

Dütschke, E., U. Schneider, and A. Peters. 2013. Who will use electric vehicles? *Working Paper Sustainability and Innovation*(6).

Feurer, R. 2012. The future of sustainable mobility, 23. BMW Group.

Franke, T., and J.F. Krems. 2013. Understanding charging behaviour of electric vehicle users. *Transportation Research Part F: Traffic Psychology and Behaviour* 21: 15.

Ghosh, T. 2014. Economic and environmental impacts of plug-in hybrids and battery electric vehicles: A comparative analysis. Master's Projects, University of San Francisco.

Hofer, C. 2013. Charging for e-mobility. MSc Computer Science, University of Twente, Twente.

Korean Automobile Industry Annual Report 2013. 2014. Korea Automobile Manufacturers Association (KAMA).

Kraus, W. 2013. Towards e-mobility: The challenges ahead, 17. Brussels: Federation Internationale de L'Automobile.

Matthies, G., K. Stricker, and J. Traenckner. 2014. The e-mobility era: Winning the race for electric cars. Bain & Company.

Moss, M. L., and H. O'Neill. 2012. Urban mobility in the 21st century, 28). NYU Rudin Center for Transportation Policy and Appleseed.

Reithofer, N. 2014a. Analyst and investor conference. BMW Group.

Reithofer, N. 2014b. Sustainable value report 2013. BMW Group.

Richter, V. (2014). Reinventing the motor car: The city mobility concept of BMW i (pp. 44). Melbourne: BMW Group.

Russell, M., D.J. Buchko, and J. Arguelles. 2013. The BMW i3: The ultimate driving machine in a new era of mobility, 18. BMW Group.

Sattig, M. 2014. The future of sustainable mobility: Leading e-mobility into a new era, 25. BMW Group.

Schlick, T., and W. Bernhart. (2013). The development of E-mobility in the global markets (pp. 21). Frankfurt: Roland Berger Stragey Consultants and Forschungsgesellschaft Kraftfahrwesen mbH.

Srečanje, O. 2013. Taking the next step: Sustainable production and products, 20. BMW Group.

Storandt, S., and S. Funke. 2013. Enabling e-mobility: Facility location for battery loading stations. Association for the Advancement of Artificial Intelligence.

Tagscherer, U., and R. Frietsch. 2014. E-mobility in China: Chance or daydream? *Innovation Systems and Policy Analysis, Fraunhofer ISI Discussion Paper* 40: 33.

Wessing, T. 2010. *Future eMobility study*. Technische Universität München.

Chapter 7
Hyundai Motor Company Case – Fostering Social Enterprises

Jay Hyuk Rhee, René Bohnsack, and Sam Lee

Backgrounds and Examples of the CSR Trend

CSR Trend in Auto Industry

In recent days, the automotive industry has been facing massive changes. Today the automotive industry contributes to the prosperity of developing countries, in the same way it supported the economic prosperity of developed countries.

While the automotive industry has been viewed positively for creating jobs, it has also been criticized for damaging the environment. Recent criticisms are also due to the changing environment in which the automotive industry operates. An IBM trend report points out that firms in the automotive industry should, for instance, improve the utilization of resources in order to minimize waste.

Moreover, the needs and wants of consumers are changing rapidly. Automobiles, simply as a means of transportation, may no longer satisfy the needs of consumers in the future. Ethical consumers may also demand automakers to tackle sustainability issues such as pollution.

According to the IBM Automotive 2020 Global Report, "Corporate Social Responsibility" will become one of the most important issues in the industry. The report suggests, for example, that "environmental friendliness" will become a more

J.H. Rhee (✉)
International Business/Strategy, Korea University Business School, Seoul, South Korea
e-mail: jayrhee@korea.ac.kr

R. Bohnsack
Catolica Lisbon, Lisbon, Portugal
e-mail: r.bohnsack@ucp.pt

S. Lee
InnoCSR, Shanghai, China
e-mail: office.china@csr-company.com

G. Lenssen et al. (eds.), *The Role of Corporate Sustainability in Asian Development*, Advances in Business Ethics Research 7,
DOI 10.1007/978-3-319-45160-2_7

important determinant of automobile purchase than traditional ones as "cost," "credibility," and "fuel efficiency." In a similar vein, a Boston Consulting Group's report on the future trend in the automotive industry emphasizes the increasing importance of automotive companies' socially and environmentally responsible activities.

So far, most studies on the sustainability of the car industry have focused on the environmental activities and less so on social activities (Orsato and Wells 2007). Studies were mainly concerned with environmentally friendly technologies (Frenken et al. 2004), strategies of car manufacturers to bring low-emission vehicles to the market (Bohnsack et al. 2014; Pinkse et al. 2014) or the health impact of cars such as mortality caused by pollution (Samet et al. 2000). This points to the fact that issues such as community impacts of car manufacturers have been neglected so far and indicates a need to have insight in social activities of car manufacturers. Thus, before examining social activities at HMC specifically, we first provide a brief account on the social activities of major car companies, focusing on German car manufacturers since they have been often front-runners with regard to corporate social responsibility activities.

Social Activities of German Car Manufacturers

We analyzed sustainability reports of German car companies to get insights into what their main focus is with regard to social engagement close to production sites and local communities in general, examining which initiatives they engage in and what key performance metrics they use (Table 7.1).

Generally, the three companies take a pro-active approach to their CSR activities, yet they have slightly different perspectives. BMW views corporate citizenship as an integral part of its business. In doing so, it focuses on areas in which it can use its core expertise.[1] Daimler equally wants to take its role as a 'good neighbor' using its competences to serve society.[2] For Volkswagen, corporate social responsibility refers to taking civic responsibility beyond factory gates.[3]

One area that the three companies have in common is their programs to teach traffic safety. Daimler has the MobileKids program, Volkswagen the ParquePolo program and BMW the Junior Campus. These programs are often initiated locally at production sites but have also been transferred internationally. For instance, BMW's Junior Campus is also active in South Korea in which BMW seeks to educate traffic safety to less privileged children in rural areas. Other general social

[1] BMW Group: Sustainable value report 2013

[2] Daimler Sustainability Report 2013

[3] Volkswagen Group Sustainability Report 2013

Table 7.1 Selected social activities of German car manufacturers derived from their sustainability report

	BMW	Daimler	VW
General areas and Focus	Traffic safety	Traffic safety	Traffic safety
	Corporate volunteering	Corporate volunteering	Corporate volunteering
	Intercultural innovation	Science, technology and environment	Future oriented, education and social projects
	Social inclusion	Art and culture	
		Societal &political dialog	
Examples	Support educational programs in South Africa	Daimler and Benz foundation	"Parque Polo", an area providing road safety training through play for children and sited in the grounds of the Volkswagen Navarra plant
	Support programs that teach democratic values in Jewish and Arab schools	Mobile kids – teach children safety and future mobility	
		You Daimler knowledge community – improve access to education	
		Sponsoring cultural festival eg Art Beijing	
	BMW warm heart fund in China – financial and educational support to outstanding students from less privileged families	Corporate volunteering	The "Volkswagen pro Ehrenamt" (Volkswagen supports volunteering) initiative is aimed both at current employees and their partners and retirees. It operates as a clearing house, linking community initiatives looking for volunteers with Volkswagen staff wanting to help their society.
KPI	Donations/ contributions in kind	Donations and sponsorship in:	Expenditures on donations
	Community investments	Education	Number of employee volunteers
	Commercial activities	Science/technology/ environment	
		Political dialog	
		Art and culture	
		Charity/community	
		Number of employee volunteers	
Donations	8.5 m Euro (28.9 m Euro)	60 m Euro for NGOs, sponsorship of social programs	19 m Euro

BMW distinguishes between donations and expenditure on including community investments which includes donations as well topics that all companies address are volunteering programs and donations. These are also mostly used as key performance indicators. By far the largest donator is Daimler (60 m Euros) whilst Volkswagen donates the least (19 m Euros).

However, within the largely homogenous approach to use core competences to serve local communities, practice donating and volunteering programs, the companies diverge on their focuses in social activities. For instance, BMW focuses on social inclusion, Daimler on fostering science and technology and Volkswagen on future oriented projects. In doing so, BMW has international programs for social inclusion (e.g. teaching democratic values), Daimler initiated a Daimler KnowledgeCommunity and Volkswagen started a volunteering community ('Volkswagen pro Ehrenamt').

Automobile Industry in Korea – Comparing with German Automobile Industry

From the previous section it becomes evident that German car companies have fairly similar approaches to social citizenship, supporting corporate donations and volunteering but also focusing on more specialized areas. All of these with a focus of matching core competences with social programs. However, how do German car manufacturers' activities differ from those of Korean car manufacturers?

A useful way to compare social responsibility activities is the materiality matrix. This matrix is based on a company exercise that compares important issues rated by stakeholders with those of the company. We have chosen Hyundai Motors, Korea's biggest automaker, as the representative of Korean automobile industry.

In Table 7.2 we compare the top five topics, which score highest in importance for the company stakeholders and the company itself for four respective companies.

This matrix is used by companies, amongst other goals, to determine the focus of their CSR activities. Interestingly, the matrix by Volkswagen identifies traffic safety to be the second highest category in terms of stakeholder expectations, but rates it fourth lowest among 16 topics regarding the importance for the company (e.g. most important is its attractiveness as an employer).

What becomes evident is that for German companies, environmental protection is one of the most important issues that they have in common. This is not part of the five most highly ranked issues of Hyundai (in fact it ranks after rank 13). What Hyundai has equally in common with Daimler and Volkswagen is their focus on nurturing talented employees. Furthermore, what Hyundai does focus on and the German companies not at all, are ethical issues such as in assessing suppliers or transparent management. Closest comes Daimler with its focus on integrity. This highlights that there are differences between German and South Korean car companies.

Table 7.2 Top materiality topics for the companies in 2012/2013 – comparing BMW, Daimler, VW with Hyundai

BMW	Daimler	VW	Hyundai (Hyundai Materiality Analysis)
CO2 emissions and climate change	*Customer satisfaction*	*Attractiveness as an employer*	*Program for systematic talent nurturing*
Alternative drivetrain technologies	*Fuel consumption and CO2 emissions*	*Environmental and climate protection*	*Implementation of transparent ethical management*
Efficient use of resources and recycling management	*Integrity*	*Economic stability*	*Fairness and transparency in suppliers evaluation*
Energy supply/ renewable energy	*Employer attractiveness*	*Customer satisfaction*	*Strengthening of quality competence*
		Resource-efficient products and production	*Expansion of eco-friendly R&D*

Social Demands of Social Enterprise in Korea

In this part, the way how car companies in South Korea engages in CSR activities is described in more detail. After the financial crisis in Korea in 1997, there was a temporary increase in job openings due to the government's financial aid. However, the supply of jobs could not be stabilized at the required level, thus criticisms and doubts on governmental aid's effectiveness have been raised repeatedly.

In the 2000s, a discussion on implementing a European like system regarding social enterprises started. Social enterprises were expected to increase the job supply combined with economic growth. Implementing a social enterprise system as an effective tool to create jobs and to offer social service of fair quality, utilizing NPOs (Non Profit Organizations), were then fostered.

In 2007, the government brought the Social Enterprise Promotion Act into effect, based on a master plan of the government to encourage and systematically support social enterprises. With the implementation of the Act for the period 2008–2012, the number of social enterprises significantly increased from 50 to 774; the number of employees in those enterprise from 1403 to 18,689. The government took further steps in December 2012 and set up four major objectives that aim to (1) strengthen the sustainability of social enterprises, (2) provide customized supporting systems for social enterprises, (3) expand the roles and responsibilities of social enterprises, and (4) strengthen the partnership between the local community and social enterprises. With this plan, the government expects to foster 3000 social enterprises during the period 2013–2017.

Social enterprises seek to match social services with public needs, offering stable jobs to socially vulnerable groups while utilizing local resources and contributing to economic growth. In recent years, the role of social enterprises has gained more attention in Korea due to the continuing trend of low economic growth accompanied by low employment but also an increased demand of services in the context of an aging society. Coupled with such trend, the government policies seem to have led many Korean firms to pay more attention to social entrepreneurship.

Hyundai Motor Company in the Korean Automobile Industry

Hyundai Motor Company (HMC)

Founded in 1967 in South Korea, Hyundai Motor Company (HMC) is now the 8th biggest multinational automobile company in the world. HMC, along with its 32.8% owned subsidiary, Kia Motors, together comprise the Hyundai Motor Group (HMG), which is the world's 5th largest automaker. In 2013, HMC and HMG sold over 4.7 million and 7.5 million vehicles worldwide, respectively. HMC has 104,731 employees worldwide, and its vehicles are being sold throughout 193 countries.

History

The history of HMC dates back to the 1940s, which was the Japanese colonial era in Korea. The founder of HMC, Jung Joo-Young, acquired a car maintenance shop named "Ado Service Garage". He made the business prosper by charging high prices to customers by offering quick and perfect service. However, his business faced a critical downturn after a fire accident and war. The Japanese government commenced a policy to restrict Korean corporations, and Ado Service was coercively merged by a Japanese company.

In 1946, Jung got his business back and renamed it as "Hyundai Motor Service Center". It mainly dealt with subcontracts from the U.S. Army.

The official history of HMC begins in 1967, when Jung started to engage in the automotive business. In December 1967, Jung's brother, Jung Se Yeong founded an automotive company and established a business collaboration with Ford Motor Company. At the early stage, the company manufactured its first model, the "Cortina".

In 1973, separated from Ford, the Jung brothers started to develop their own automobile model. They hired George Turnbull, a former vice president of British Leyland, and made a technology contract for important automotive parts such as engines, accelerators, and transmission. In 1976, HMC started to produce the first Korean automobile, the "Pony", and soon produced the model "Sonata", the beginning of a journey to becoming the biggest automaker in Korea.

HMC expanded its business into North America in the 1980s, and acquired KIA Motors Company and Asia Motors Company in 1998. Two years later, HMC formed HMG, separating from the Hyundai Group together with other 9 affiliates.

Global Sales

In 1990, HMC marked for the first time one million units in global sales. It was largely due to the popular 'Excel' model, which gained its reputation by cost efficiency. In 1996, the number of automobiles sold grew to over two million. Due to its merger with KIA Motors in 1998, the degree of sales increase got even bigger. In 2004, HMG marked 3.14 million unit sales. The number of products sold surged afterwards, with 4.18 million in 2008, 5.74 million in 2010 and 6.6 million in 2011. In 2014, HMG expects to reach eight million unit sales worldwide.

Originally, HMG had set the goal at 7.86 million due to global economic recession, weak Yen and Japanese automakers' low price strategy. However, due to sales increases in the BRIC countries (10.5% in China, 8% in India, and 7.2% in Brazil, compared to last year), the annual sales are expected to exceed the target (Fig. 7.1).

Geographical Operations

Currently HMC has seven overseas plants in the United States, India, China, Turkey, the Czech Republic, Russia and Brazil. Employing over 78,000 people worldwide, HMC is implementing a new global policy of localization. That policy includes the adjustment of brand positioning strategies in order to better appeal to unique local demands (Rhee and Kim 2014).

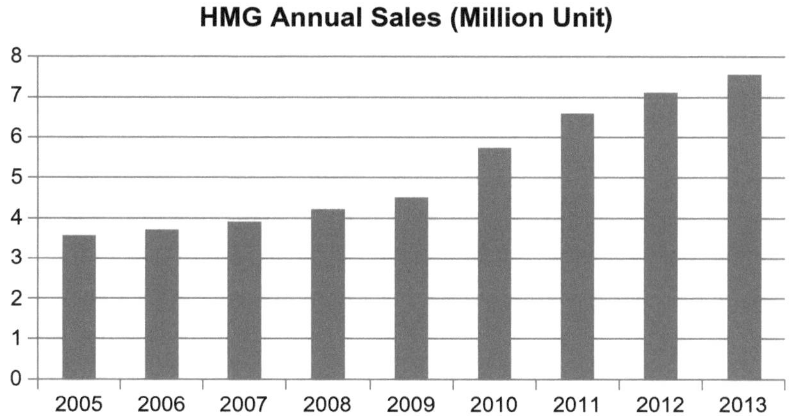

Fig. 7.1 HMG annual sales (Korea Automobile Manufacturers Association(KAMA), Monthly Automobile Statistics(2014))

HMC and CSR

After by the South Korean government passed the Social Enterprise Promotion Act in 2007, HMC established a CSR Committee to promote its CSR activities more effectively in 2008. The CSR Committee is responsible for activities in three core areas including environmental management, trust-based management and social contribution.

Some of HMC's CSR objectives are to create jobs through start-up assistance, to foster global youth leaders, and to alleviate social polarization by improving the quality of life of socially vulnerable groups (Fig. 7.2).

From 2004 to 2007, the annual budget for social contribution activities of HMC was rather fixed under 40 billion KRW. It then increased from 29 billion KRW in 2007, 50 billion KRW in 2008, and to 70 billion KRW in 2009. Recently it reached 74 billion KRW.

CSR Philosophy

Under its slogan, "Realizing the Dreams and Aspirations of Humankind through Creative Thinking and Rising to New Challenges," HMC has made strong commitment to its philosophy and social responsibilities (Fig. 7.3).

Hyundai divides its CSR activities into economic, social, and environmental responsibilities. The economic responsibilities, serving the fundamental basis of

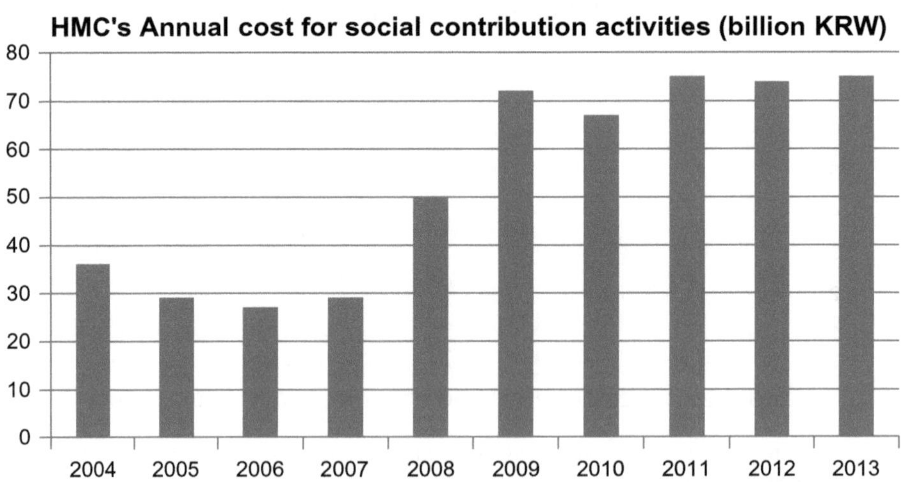

Fig. 7.2 HMC's annual cost for social contribution activities (HMC Social Contribution Report (2014))

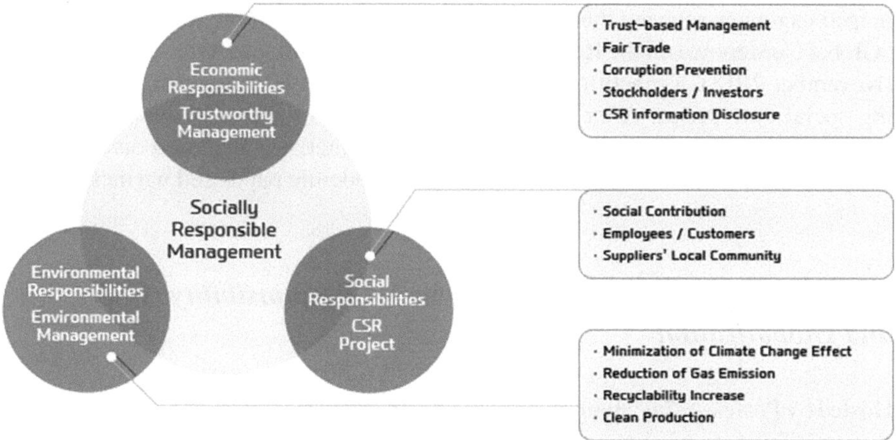

Fig. 7.3 The conceptual area of HMC's CSR strategy (http://worldwide.hyundai.com/WW/Corporate/CorporateInformation/CSR/index.html, Accessed Dec 2014)

any profit-making organization, include the needs to create jobs and economic values. Such responsibilities are fulfilled by trustworthy management. HMC also fulfills its social responsibilities through its various CSR campaigns and programs catering to the needs of its stakeholders, including the socially vulnerable groups at large, employees, customers, business partners and local communities. The company has also been implementing environmental pollution management systems in order to fulfill the environmental responsibilities.

CSR Structure

HMC is engaged in many CSR projects including both the '4 Moves' and the '4 main projects'; the former reflects its corporate enterprise, while the latter includes fostering social enterprises, social contribution on culture and art, global social contribution, and its 18 affiliates' projects. Moreover, HMC is collaborating with the Jung Mong-koo foundation to practice value of sharing. In this case, we will mainly focus on HMC's social enterprise fostering project.

Research Method

In the process of writing this case, we had mainly relied on the feedback from the two conferences on CSR, conducted interviews with practitioners of HMC, and reviewed documents of HMC corporate reports. The first conference, "Global Conference on Corporate Social Responsibility and Globalization" (April 2012), helped us review relevant theories and literature on CSR. At the second conference, "Global Conference on the Role of Corporate Sustainability in Asia's Development" (November 2013), a practitioner from HMC presented HMC's practices on fostering social enterprises. After finalizing the contents of the case, we conducted a series of face-to-face interviews with HMC personnel. We also used diverse information from corporate reports as well as from academic paper and business media.

Global Conference on Corporate Social Responsibility (CSR) and Globalization

Hosted by Professor Jay Hyuk Rhee at the Asian Business Center at Korea University Business School and the Academy of Business in Society, this conference mainly focused on the theme of "Corporate Social Responsibility (CSR) and Globalization" with three sessions: (1) CSR as a Global Strategy, (2) CSR in Practice, and (3) How to Assess and Improve CSR.

The conference brought together distinguished experts and professionals in CSR. It provided a great opportunity for researchers to establish relationships with CSR experts and to deepen their potential commitment to the professional development in the field of CSR.

Global Conference on the Role of Corporate Sustainability in Asia's Development

This conference hosted by Professor Jay Hyuk Rhee at the Asian Business Center at Korea University Business School and the Academy of Business in Society invited interested academics and practitioners to share and discuss their views on the role of corporate sustainability in the development of the region. Attended at the conference were a total of 14 scholars from 10 countries, Australia, Bangladeshi, Canada, China, Germany, India, Korea, Malaysia, UK, and U.S. as well as 7 practitioners representing the ICT industry (Intel, Lenovo, Samsung Electronics, ZTE) and the automobile industry (BMW, Hyundai Motor Company, and Mahindra & Mahindra).

The aim of the second conference was to discuss current challenges and share the best practices that have contributed in the development of Asia or that can further be applied in order to expand the impact their positive impact in Asia. HMC, BMW and Mahindra & Mahindra participated in the conference to hold sessions about the sustainability issue in the automobile industry.

1. BMW
   : BMW i Story – Revolutionizing Sustainable Mobility
   Presenter: Inchul Cho (Head of Representative Office, Corporate and Government Affairs, BMW Group Korea)

2. Hyundai Motor Company
   : Fostering social enterprises
   Presenter: Hyeon-sook Heo (Deputy General Manager, Hyundai Motor Company)
3. Mahindra & Mahindra
   : Mainstreaming Sustainability in Business through knowledge building
   Presenter: Beroz Rumie Gazdar (Senior Vice President-Group Sustainability, Mahindra & Mahindra Limited)

Constant Interviews with Extensive Research

During the data collection process for this case, interviews with practitioners in HMC were conducted on a regular base. They included: Deputy General Manager Hyeon-sook Heo and Manager Jae Ho Choi at Corporate Responsibility Team of Hyundai Motor Group.

Most information and exhibits stated in this case came from the internal information of HMC.

HMC Corporate Reports, Website and Media Coverage

For statistics and some exhibits, information on the official HMC website, CSR website, HMC annual reports and HMC's sustainability reports were used. A few quotations and information were found from major media in Korea, such as Dong-A Daily, The Korea Economic Daily, etc.

HMC and Social Enterprises

HMC has been supporting social enterprises in many different business fields such as services, manufacturing and retail. It has also been providing job opportunities to the socially underprivileged. In line with the government objectives, the strategic focus of the CSR activities of HMC is based on its CSR philosophy which puts more value on social enterprises – HMC had a judgment that social enterprises are the fundamental solutions for social issues since they create economic, social values through their business operations and welfare system rather than the traditional form of unilateral philanthropy.

HMC has constantly founded and supported several social enterprises. The social enterprises include (1) Easy Life, the first care service social enterprise for the elderly and the handicapped in Korea (2) Easy Move, the first manufacture-based

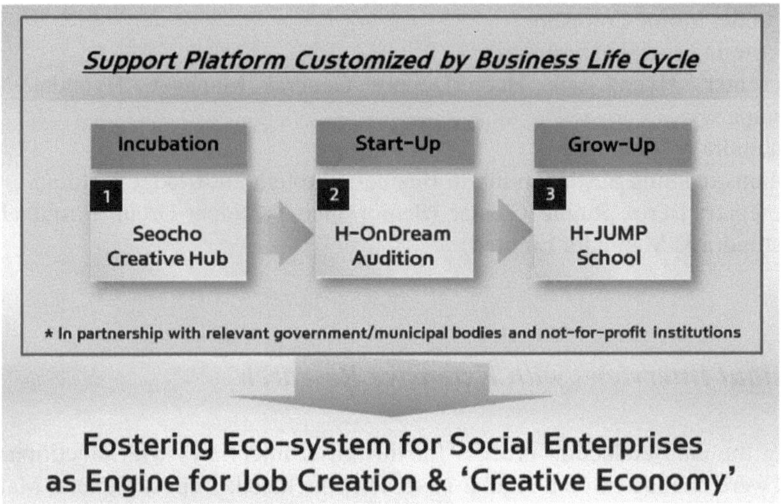

Fig. 7.4 HMC's social enterprise support projects life cycle (2013 Global CSR Conference, HMC)

self-supporting social enterprise producing the handicapped aids in Korea. Moreover, HMC has been conducting several projects to support social readaptation of female North Korean defectors and juvenile delinquents (Fig. 7.4).

Besides, due to its corporate philosophy that the core competencies of a social enterprise are derived from its human resources, HMC has been concentrating on investments to foster social entrepreneurs. HMC aims to support 150 social enterprises and create 1500 jobs within 5 years from 2013 through 'Seocho Creative Hub', a fostering center for prep social entrepreneurs, and 'H-OnDream Audition', a program to discover and support youth social entrepreneurs. HMC has been also conducting H-Jump School, connecting social enterprises with young students.

Fostering Representative Social Enterprises

Starting in 2008, HMC has been fostering representative social enterprises in service, manufacturing and retail business areas and supporting Easy Life and Easy Move, which are described in more detail in the following.

Easy Life

In 2008, HMC founded 'Easy Life', a social enterprise that deals with care service for the elderly and the handicapped. With Easy Life, HMC has created 700 jobs for unemployed women. The recipients of the high-level service offered by Easy Life have reached about 340 thousands (in 2013). Easy Life operates four elderly health care facilities and twelve social franchises for home care service.

Easy Life recognizes the accelerating phenomenon of aging society as a serious social problem, which society as a whole should be responsible of. With high-quality care service, Easy Life is elevating the elderly welfare, hiring more than 60% of the 700 employees from unemployed women older than 50.

Easy Move

In August of 2010, HMC launched Korea's first social enterprise named Easy Move in collaboration with Gyeong-Gi province, and has been supporting its activities ever since. The company specializes in manufacturing and distributing auxiliary devices and rehabilitation-assisting equipment for people and seniors with mobility-affecting disabilities. To eliminate some preconception that goods manufactured by social enterprises have poor quality, Easy Move has 'GMP' qualification[4] from Korea Food & Drug Administration and continues to enhance the quality of its products and services by developing and localizing advanced technologies. In 2012, Easy Move marked 3.2 billion KRW sales. The company also spends over 2/3 of its operating income on promoting valuable social causes, while steadily increasing the number of jobs for members of unemployment-prone groups.

HMC and Youth Social Enterprises

HMC has made a fostering platform to support young people with innovative ideas to establish social ventures, in order to help them to create jobs and solve social problems. There are two pillars for this project: A. 'Seocho Creative Hub' to discover and foster pre-social entrepreneurs and B.'H-OnDream Audition' to fully support young social entrepreneurs. Through this project, HMC initially planned to foster 200 social enterprises and create 1500 youth jobs within 5 years.

---

[4]A GMP(Good Manufacturing Practice) is a system for ensuring that products are consistently produced and controlled according to quality standards. It is designed to minimize the risks involved in any pharmaceutical production that cannot be eliminated through testing the final product. GMP covers all aspects of production from the starting materials, premises and equipment to the training and personal hygiene of staff.

Two distinctive features of HMC's assistance project for start-ups are: (1) Unlike existing programs, HMC implemented an audition and matchmaking system between new entrepreneurs and successful social entrepreneurs, enabling mentoring and follow-up support. (2) HMC used its foundations and networks to offer financial, educational support and advisories to new entrepreneurs.

Eocho Creative Hub

Motivation

As youth unemployment problems aggravated, job-hunting became less effective than before. Social ventures came to the foreground as an innovative way to solve both the youth unemployment problem and social issues through job-creation. With the social enterprise-expert company Seeds, HMC planned a 'local complex center' which will serve two purposes - supporting start-ups and local services- at one place. During the first phase, the local government of Seocho Gu, an administrative district in which HMC's headquarter was located, suggested HMC to make better use of 'Simsan Memorial Hall'. Leasing partial space of the Simsan Memorial Hall in cooperation with Seeds and Seocho Gu Office, HMC finally opened 'Seocho Creative Hub' in October, 2011.

Main Contents

The program contents of Seocho Creative Hub are mainly divided into 2 parts: 'Incubating youth social enterprises' and 'Offering education & culture programs' to local citizens. The former program selects young prep-entrepreneurs and provides them with offices, business development expenses, business mentoring, etc. It is now being operated in association with Korea Social Enterprise Promotion Agency under the Ministry of Employment and Labor.

Another main program, the education & culture program, is offered by the youth entrepreneurs nurtured in Seocho Creative Hub. They offer educational and cultural service to local citizens. Unlike typical local cultural programs, it has created a daily contact point between local residents and social entrepreneurs.

Cooperation Model

Seocho Creative Hub is a good practice of cooperation between a big corporation and a Non-Profit Organization (NPO). HMC and Seeds, with its expertise in social enterprise fostering, made a joint planning from the first phase. Moreover, in the process of operation, the two bodies actively communicate with each other to cope with problems that occur during new business projects.

Seocho Creative Hub was organized by a collaboration of three different sectors – a company (HMC), an expert team (Seeds) and a governmental body (Seocho-Gu Office) with their resource and expertise combined altogether. Such a cooperation model is distinctive in social entrepreneurship field, since it maximizes the effectiveness through the division of roles.

Key Success Factors

Seocho Creative Hub is the first 'start-up fostering center complex', a combined form of a start-up support facility and local educational and cultural service function in Korea. Such formation enables diffusion of innovation through fostering youth social enterprises, comprising one of the important success factors of Seocho Creative Hub.

Another Key Success Factor (KSF) is fortified cooperation among different sectors. HMC has maintained close cooperation with the expert team, Seeds and Seocho-Gu office. Such a cooperation model occasionally functioned as a barrier to prompt decisions. However, it eventually worked as a positive influence to the whole project since HMC could mediate different visions of each sectors. In spite of the fact that such project has never been implemented before in Korea, resources of each sectors combined with Seed's expertise made the whole project very successful.

HMC's long-term support to the project was also one of the KSFs. Because most CSR activities in Korea are short-term based, long-term projects that do not provide tangible results within a year are not easy to get corporate support. Recognizing the social significance of the project, however, HMC has supported Seocho Creative Hub project from long-term perspective. Due to such steady support of HMC, Seocho Creative Hub could easily be connected with other government and civil organizations, literally functioning as a 'hub' for social enterprises and other innovative activities.

H-OnDream Audition

Motivation

'Youth Social Enterprise Initiative' was initiated to increase youth job creation and realize social values by fostering youth social entrepreneurs. This project has been implemented by Seocho Creative Hub, Seeds and Korea Social Enterprise Promotion Agency. Selected youth teams get support to establish their own social enterprises, moving into an incubating center[5] like Seocho Creativity hub for a year.

[5] In 2011, there were 20 incubating centers to support prep-social entrepreneurs.

However, there were concerns that such youth social entrepreneurs (about 320 teams annually) would need follow-up support in order to stabilize themselves in real-life situations. Thus, HMC decided to devise a follow-up program with Seeds to help young entrepreneurs to get settled. In the process of planning, Korea Mecenat Association[6] joined the project to support social enterprises related with culture and art. For the financial support, Chung Mong-koo Foundation took the role.

Main Contents

H-OnDream Audition selects 30 start-up teams from hundreds of entrepreneurs and prep-entrepreneurs participating in 'Social Enterprise Promotion Project' or Ministry of Employment and Labor's 'Creative Campus Project', and supports their business development. The recipients of the assistance program would be categorized into 2 groups: A. 'Launching group', which has already achieved some progress and thus receives only support funding, or B. 'Incubating group', which participates in the whole follow-up program.

The youth entrepreneurs, who have already passed the document screening, go through the preliminaries with industrial classification, and qualifiers are to compete in the final. The winning teams are qualified for a 1-year program for support, and successful teams are granted as H-OnDream Fellows with additional financial support.

Cooperation Model

One of the most distinctive features of H-OnDream Audition's cooperation model is that the cooperation between big corporations and a professional organizations has been the foundation of the project. Such cooperation is very significant since it was devised as a voluntary complement to limitation of governmental support on start-ups. Therefore, it was necessary to build up a cooperation system to coordinate interests of different groups – fostering institutes, government, social enterprise participants etc.

HMC and Seeds have proceeded the project through a 'dualized consultation system'. They organized a H-OnDream bureau and let routine decisions be made by HMC, Chung Mong-koo Foundation, Seeds, and Korea Mecenat Association, while significant decisions were made by the main council. The system enabled to gather all stakeholders' opinions without delaying routine decision-making.

[6] Korea Mecenat Association was founded in 1994, with the purpose of promoting social contribution by supporting culture and art.

Key Success Factor

Unlike other similar start-up assistance projects, H-OnDream Audition was initially made by a conglomerate's proactive response to the needs of young social entrepreneurs. As a result, the time spent between the planning stage and implementing stage was relatively short, and the effect of the project could be maximized. Also, active utilization of private expertise was a factor of success.

As mentioned above, a dualized decision-making process was another KSF. Since the project covered exhaustive number of stakeholders, dualized decision-making process could have delayed the process. However, H-OnDream Audition prevented the delay by structuring a central organization (joint bureau), while cooperating with several different institutions at the same time. In addition, HMC selected representative institutes from each region for effective division of roles.

Another KSF of H-OnDream Audition is that the project was planned in a long-term perspective from the beginning, in common with the case of Seocho Creative Hub. In fact, Chung Mong-koo Foundation's financial support is a 5-year plan. Such long-term vision is expected to contribute to lead better achievement of participants than other similar start-up assistance programs.

Moving Forward: H-Jump School

H-Jump School is an advanced concept program operated by HMC, Seoul Scholarship Foundation and Jump, a social venture. H-Jump School aims to offer benefits to all participants by applying a different approach from current scholarship programs. Public-Private Partnership, cooperation among main agents of government institutions, constant high-quality education and beneficiary-oriented implementation enabled H-Jump School to become a new solution.

Excellent students, passionate about resolving social issues, are selected and fostered as leaders with volunteer spirit and abundant knowledge in humanities. The students, selected through a systematic screening process, become tutors and give private lessons and mentoring sessions for a year to underprivileged adolescents including those from multicultural or low-income families. H-Jump School offers university student tutors scholarships, overseas training programs, special lectures on humanities, etc.

From 2013 to 2017, H-Jump School aims to foster 500 university student tutors, to offer tutoring and mentoring for 2000 underprivileged adolescents and to open mentoring sessions conducted by 100 career men including 30 from HMC.

Program Structure

Under H-Jump School program, underprivileged adolescents get tutoring and mentoring by university student tutors on a regular basis (12 h a week) for a year. University student mentors get career consultation from mentors, consisting of 100

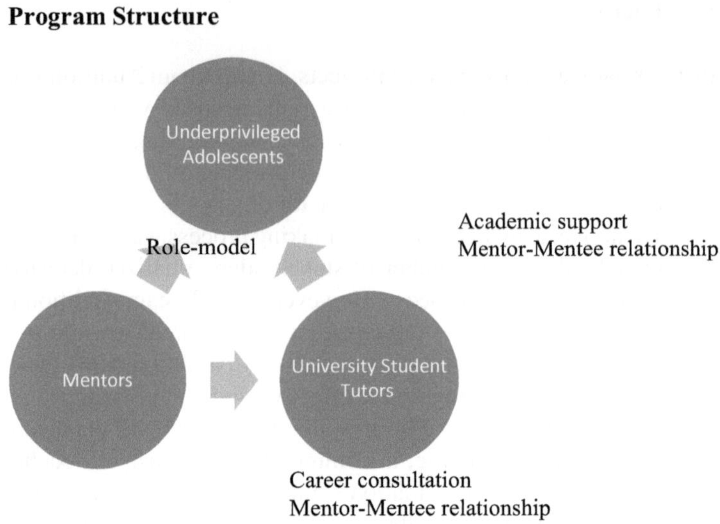

Fig. 7.5 H-Jump school structure (H-Jump School official website, http://h-jumpschool.kr/, Accessed Dec 2014)

career men in their 20s to 40s from diverse professional areas including professors, company executives and social entrepreneurs. The main structure of the program is to make a virtuous cycle that the adolescents become the tutors, and the tutors become the mentors after the program (Fig. 7.5).

Key Success Factors

H-Jump School suggests a creative Public-Private Partnership model of 'cultivating talented university students combined with educational gap settlement by corporate social contribution activities collaborating with local government and social enterprises'. Unlike temporary voluntary activities which are usually provider-oriented, H-Jump School planned the program beneficiary-oriented by making university student tutors offer regular tutoring and mentoring customized with the adolescent's academic level.

Also, H-Jump School offers abundant and diverse benefits to university student tutors by cooperating with local government and social enterprises. For example, tutors that have shown excellent performance will be offered opportunities to visit Ivy League universities in the United States, to participate in Happy Move Global Youth Volunteers[7], etc.

[7] Established in 2008, Happy Move Global Youth Volunteers is one of the representative CSR activities of HMC. Every year, 1,000 university students are provided with opportunity to volunteer in oversees countries during vacations.

Table 7.3 Numerical results of HMC social venture fostering projects (HMG 2013 CSR Handbook)

Project title	No. of Participants	No.(%) of success	Employment	Sales (Billion KRW)
Seocho creative hub	90	43(48%)	334	1.36
H-OnDream audition	60	60(88%)	347	4.4
Sum	150	103(64%)	681	5.76

Overall Summary and Evaluation

The features of social enterprise-fostering business model of HMC are as below. First, HMC is supporting social enterprises utilizing the cooperative relations between HMC's corporate sociocultural team and external expert groups. Second, HMC is proceeding a sustainable social enterprise business model by fostering representative, job-creating social enterprises such as Easy Life and Easy Move in service, manufacturing and retail sectors. Third, HMC seeks to build a platform to support youth social entrepreneurs, discovered through H-OnDream Audition, from educating to supporting actual business foundations. HMC includes social entrepreneurs fostered through H-OnDream Auditionin H-OnDream Fellows and develops new strategic social contribution projects through collaborating with Jung Mong-koo foundation.

Performance

Regarding HMC's projects of fostering social enterprises, numerical and exemplary performance indicators are described as below (Table 7.3).

Results

From 2011 to 2013, Seocho Creative Hub created 334 jobs, supporting 43 successful social enterprises out of 90 participants, to generate 1.36 billion KRW in sales.

From 2012 to 2013, HMC fostered 60 social enterprises and created 347 jobs in total through H-OnDream Audition, the youth social entrepreneur fostering project. Every year, HMC is going to select 30 teams (150 teams for 5 years), constantly and systematically support them by offering them start-up education, consulting, maximum of 150 million KRW funding per team, mentoring by successful social entrepreneur, etc. Through such a program, HMC aims to cultivate young future leaders that lead social innovation.

Table 7.4 Examples of H-OnDream fellows 1st

Category	Name	Contents	Employment	Sales (KRW)
Culture & art	Joyul	B-boy contents production	12	110 million
	Learning to learn	Local development through art activities	4	65 million
IT	Tumble buck	Online social funding platform for culture creators	5	30 million
	My real rrip	Customized travel platform	8	200 million

The initial goal was to create 600 jobs but H-OnDream Audition has already created 370 jobs in the first 2 years. This resulted in the initiative receiving a good evaluation for employment creation.

Exemplary Cases

As described in the previous sections, HMC's social enterprises have been involved in diverse categories including culture, education, etc. In March of 2014, the 3rd H-OnDream Audition was held for 300 teams nationwide that completed youth social entrepreneur promotion program hosted by the Ministry of Employment and Labor. The jury, consisting of experts from diverse fields, conducted screening, preliminaries and the final and selected 15 incubating groups and 15 developing groups.

Teams with various business plans such as vitalizing traditional markets, developing gukak (Korean classical music) contents, were selected. The grand prize was awarded to 'AUD social cooperative' team, which devised a business plan for a real-time, mobile subtitle platform for the hearing-impaired (Table 7.4).

As mentioned above, successful teams among those selected in H-OnDream Audition are appointed as H-OnDream Fellows, and become mentors for prep social entrepreneurs. Joyul was launched by Yong-Nam Song, a world class B-Boy champion, who used to be a juvenile delinquent. He secured about 1000 m^2 space in Bucheon, which is on procedure of a project to make it civil culture art educational center. My Real Trip, which invited 400 million KRW investment in 2013, built a direct transaction platform between tourist guides and tourists in 110 cities in 32 countries. After it participated in H-OnDream projects, its sales mounted up 11 times bigger.

Outlook on Future Challenges

Corporate social responsibility (CSR) has been attracting more and more attention from media. Popular business magazines such as The Economist and Business Week have produced numerous stories and articles related to CSR. Such public focus on CSR has put up a lot of challenges for numerous corporations.

Being a very visible target for policy makers and non-governmental organizations, HMC, just like other corporations, has been under heavy pressure to become more socially responsible. However, 'doing well by doing good' is not as easy as expected.

As pointed out in many studies, there is always a possibility of breakaway from a corporation's core economic purpose of maximizing shareholders' stock value because it may fail to recognize the significance of the purposes. On the other hand, there is also a perspective that regards CSR activities as a source of long-term, sustainable corporate competency.

The main issue for HMC in this case lies in sustainability. Like other CSR activities, HMC's support on fostering social enterprises is mainly targeted towards social contribution, which does not seem to explicitly link to corporate profit increase. Plus, it takes considerable time to detect and measure the increase in profit caused by CSR activities. However it seems to have been fostered by the government, which is an interesting area for further research. This also relates to the comparison with German car manufacturers, which seem to focus on environmental sustainability rather than social responsibility. This may have been driven partially by the commitment of the German government to tackle climate change.

The extent of being socially responsible may vary, ranging from the most reactive attitudes to the most proactive ones. In other words, being socially responsible can be perceived as a strategic alternative. From HMC's point of view, however, the real challenge is how to capitalize on their CSR practices more effectively. According to Becker-Olsen et al. (2006), when a corporation and social initiatives' fit well, and the corporation proactively put efforts into enhancing the 'fit', it can draw from a more positive attitude, trust and higher purchase intention from customers. Also, David et al. (2005) have verified that through CSR practices, consumers recognize a company more innovative and professional. Consumers have the tendency to think of a company with proper CSR practices as more trustworthy, and have a more positive attitude toward the company. Such attitude leads to higher purchase intention. Thus, more familiar CSR activities to consumers mean a higher potential to positively affect consumers' purchase intention.

To sum up, HMC's CSR practices on fostering social enterprises will need more customer connection to make its existing and new customers feel more familiar with the CSR practices. Such connection would help HMC to create more positive attitude among consumers, which would result in higher purchase intention. Although HMC has recently put more efforts on communicating with customers about its CSR practices, making more connection with customers in the process of the practices would be another task.

Future Plan

In order to enhance the 'fit' between a company and social initiatives, one of the most important factors is strong commitment from the upper management. Fortunately, HMC seems to express its passion toward further CSR practices. According to The Korea Economic Daily, HMC's vice chairman Chung Eui-Sun has mentioned that "Although business environment for HMC is not as favorable as before, HMC promise our customers to be more sustainable and responsible company than before."

> *"Hyundai is also endeavoring to carry out social contribution activities that contribute to creating greater shared value which will no doubt help us maintain our credibility and respect as a corporate citizen. We intend to create and share new value by providing high quality services beyond customer's expectation."*
>
> **Quote 6.1** March 2014, Chung Mong-koo, Chairman, Hyundai Motor Group[8]

HMC plans to conduct another CSR project, as a further step for its social enterprise supporting programs, connecting social enterprises with local district.

H-Village Project

HMC has recently been pushing forward a plan to revitalize a village named 'Chang-shin Dong' where the majority of the residents engage in the sewing industry. Local associations, social enterprises, the Chung Mong-koo foundation, the Korea Mecenat association and others are cooperating for the project.

About 'Chang-shin Dong'

Near Korea's biggest fashion town, Dong-dae mun area, there is a village called Chang-shin Dong, where approximately 2000 small sewing factories are located in. After 1970s' labor movement, sewing factories originally located in Dong-dae mun area scattered into nearby residential areas, and Chang-shin Dong was one of them. About 70% of the residents in Chang-shin Dong are engaged in sewing industry. The sewing factories are mostly small-sized, operated by 2–3 people who are families or acquaintances.

However, for the last 15–20 years, young people in Chang-shin Dong have increasingly adopted other professions than the traditional sewing jobs. Due to such a phenomenon, Chang-shin Dong would have to go through a lot of changes. HMC wanted to lead such changes in proactive and constructive way, and plans to seek a new development of Chang-shin Dong.

[8] HMC, Sustainability report(2014)

Key Points

1. Active participation of local residents needed
2. H-Village must be self-sustainable, not being too dependent on fund
3. The programs should be held constantly, not a one-time event
4. The project should be proceeded in cooperation with local associations

Programs

Furniture in the Streets

Furniture made with recycled materials will be installed in the streets of Chang-shin dong. The furniture would include mainly public chairs and benches that can be utilized by residents and visitors.

12 Columns

Every day, tremendous amounts of leftover pieces of cloth get dumped in Chang-shin Dong. HMC has thought that those leftovers are not just mere trash, but a sort of archive that shows local labor in small sewing factories. '12 columns' is a public art project that everyone can use. Leftover cloth will be made into cushions, and those cushions will be piled up like columns. One column will be made each month, comprising 12 columns at the end of the year.

Village Space Network

A booklet that introduces local associations in Chang-shin Dong will be published and distributed to local residents in Chang-shin Dong. It will be practically helpful to active local projects in the area.

Culture and Art Internship for Youth in Chang-shin Dong

Chang-shin Dong revitalization project will be also conducted by local youths, not just external artists or activists like other similar projects. The local youth will be paid for what they do, being motivated to do meaningful things for where they live.

Sound Scenes

This is a new version of the 'strollers of the city' project, planned and carried out by Seoul History Museum and Learning to Learn. Other projects such as 'Furniture in the streets','12 columns', 'Village space network' and 'Culture and art internship for youth' will all be connected by this project. Sounds of every corner in Chang-shin Dong will be recorded and made into a 'Sound Map' and distributed through a smart phone application. Users will be able to stroll around the area in an intriguing way using the application. The project will bring positive influence to the local commercial area.

Travel of Leftovers

Experienced master tailors of Chang-shin Dong will become teachers themselves. They will teach the participants about the process of garment making. Children will learn how to make a broche with leftover cloths and buttons. Youth and adults will experience the process of garment making, and will have an opportunity to design their own clothes.

Production of Archive Videos and Books

Each process of making videos and books will be streamed through social media to everyone. The process will be recorded in detail, so that it can be helpful to other villages in the future. At the end of the year, a documentary film of the H-village project will be made.

References

Becker-Olsen, K.L., B.A. Cudmore, and R.P. Hill. 2006. The impact of perceived corporate social responsibility on consumer behavior. *Journal of Business Research* 59(1): 46–53.

Bohnsack, R., J. Pinkse, and A. Kolk. 2014. Business models for sustainable technologies: Exploring business model evolution in the case of electric vehicles. *Research Policy* 43(2): 284–300.

Frenken, K., M. Hekkert, and P. Godfroij. 2004. R&D portfolios in environmentally friendly automotive propulsion: Variety, competition and policy implications. *Technological Forecasting and Social Change* 71(5): 485–507.

HMC 2014 Sustainability report – The road to sustainability. http://csr.hyundai.com/da_400_01.do?CSR_LOCALE_PARAM=en.

HMG. 2013. CSR handbook.

Orsato, R., and P. Wells. 2007. U-turn: The rise and demise of the automobile industry. *Journal of Cleaner Production* 15: 994–1006.

Pinkse, J., R. Bohnsack, and A. Kolk. 2014. The role of public and private protection in disruptive innovation: The automotive industry and the emergence of low-emission vehicles. *Journal of Product Innovation Management* 31(1): 43–60. doi:10.1111/jpim.12079.

Prabu, D., S. Kline, and Y. Dai. 2005. Corporate social responsibility practices, corporate identity, and purchase intention: A dual-process model. *Journal of Public Relations Research* 17(3): 291–313.

Rhee, J.H., and S. Kim. 2014. Brand positioning strategy and global competitiveness: The case of Hyundai Motor India. *International Business Review* 25(1): 59–87.
Samet, J., F. Dominici, F. Curriero, I. Coursac, and S. Zeger. 2000. Fine particulate air pollution and mortality in 20 US cities, 1987–1994. *The New England Journal of Medicine* 343(24): 1742–1749.
Seocho Creative Hub official website, http://schub.tistory.com/. Accessed Nov 2014.

Chapter 8
Mahindra & Mahindra – Mainstreaming Sustainability Through Knowledge Building

Tapan Sarker, Subhasis Ray, and Beroz Gazder

Case Overview

Mahindra & Mahindra Ltd, (hereafter referred to as the Mahindra Group or the Group), is an Indian multinational conglomerate headquartered in Mumbai. It has a presence in over 100 countries worldwide with interests across a vast number of industries. In India, it is the market leader in the manufacturing of utility vehicles and tractors. This chapter describes the steps taken by the Mahindra Group in mainstreaming sustainability in its business operations. It provides a detailed case study outlining its use of a knowledge building approach that has helped the Mahindra Group to adopt an "alternative thinking" approach to enable innovation and change by enhancing its new brand positioning 'Mahindra RISE'.

This new brand positioning provides a means by which not only the employees but also the community at large, can align actions and behaviors and 'live' the Mahindra values. The three Brand Pillars of 'Mahindra Rise' are: Accept no Limits; Alternative Thinking, and; Drive Positive Change. The values and the three Brand Pillars thus provide the impetus for an ongoing focus on sustainability within the Mahindra Group. This chapter and the case study resulted from a collaborative

T. Sarker (✉)
Griffith Centre for Sustainable Enterprise, Griffith Asia Institute, Griffith Business SchoolGriffith University, Brisbane, Australia
e-mail: tapan.sarker@griffith.edu.au

S. Ray
Centre for Resettlement, Rehabilitation and Corporate Social Responsibility, Xavier Institute of Management Bhubaneswar, Bhubaneswar, India
e-mail: sray@ximb.ac.in

B. Gazder
Group Sustainability, Mahindra & Mahindra Ltd, Maharashtra, India
e-mail: gazdar.beroz@mahindra.com

G. Lenssen et al. (eds.), *The Role of Corporate Sustainability in Asian Development*, Advances in Business Ethics Research 7,
DOI 10.1007/978-3-319-45160-2_8

effort between executives of the Mahindra Group and management scholars from leading Indian and Australian business schools.

Introduction

Chapter Outline

The Mahindra Group in India has positioned itself as a leader in sustainability in India and is also increasingly seen as a force to be reckoned with in an international context. Its approach to enmesh sustainability in all aspects of its business is an example of the use of a knowledge building, knowledge management approach to sustainability. This chapter will explicate its approach from this perspective. As such, this chapter will begin by providing in introduction to knowledge management and sustainability from a global and Indian perspective. An overview of the evolution of the sustainability agenda within the Indian context will be provided, and the link between knowledge management and sustainability from an Indian perspective will be established.

Following from this, a brief introduction of the Mahindra Group will provide context and background for the case study. The case study will then provide insights into the approach taken by the Mahindra Group. It will provide an overview regarding its initial steps to developing understandings about the potentials for sustainability and the challenges it could encounter when embarking on a sustainability path. Following from this, its initial approach and its supporting management structure will be introduced. An overview of its activities regarding raising general awareness and creating management buy-in will be provided and its Road Map to sustainability explicated. The case study will wrap up with an overview of achievements and outcomes to date and a summary of it external collaborations and partnerships.

The chapter will conclude with an explication of the Mahindra Group case study as an example of a knowledge management approach to sustainability and concluding remarks.

Knowledge Management and Sustainability

The rapid growth of technology has given rise to an economy where competitive advantage is increasingly based on an organization's ability to apply knowledge and where knowledge becomes the tangible resource (Lengnick-Hall and Lengnick-Hall 2003). Thus knowledge management is an important activity in organisations. Knowledge management revolves around deliberate, explicit, systematic and organized processes that enable the creation, dissemination, application, reviewing and updating of knowledge in order to achieve organisational objectives. Knowledge comprises of all those tangible attributes such as technology, products, processes, customers, markets etc., and intangible attributes such as insights, experiences and reasoning that enable action to be taken (Pillania 2008). However, it requires that people as architects of the knowledge creation process and the management of stakeholder relationships are placed at the center. In the knowledge management literature, it is recognized that different types of knowledge exits. Most commonly the knowledge dimensions are categorized around two different perspectives; tacit knowledge as the type of knowledge that a person inherently possesses without being conscious; and, explicit knowledge as the type of knowledge that is held consciously and is available for communication and sharing (Seow et al. 2006). Knowledge management is then also likened to the capability of an organization to be able to capture its explicit and tacit knowledge that is influenced by a number of organizational factors (Gold et al. 2001). For example, Gold et al. (2001) suggest that the ability for an organization to capture knowledge requires a knowledge infrastructure, which consists of technology, structure and culture. They also suggest that essential organizational capabilities then revolve around a knowledge process architecture, which consists of acquisition, conversion, application and protection. Following from this, Seow et al. (2006) suggest that establishing knowledge management in organisations follows a specific roadmap. Their STEPS Knowledge Management Maturity Road Map reflects the various stages in which the knowledge management process matures. It can be used to both analyze the progress made in the maturity of knowledge management processes within an organisation and inform its ongoing development. It represents a total of five stages: Start-up; Take-off; Expansion; Progressive; and, Sustainability; where the latter refers to the sustainability of the knowledge management processes within the organization. In the Start-up stage, an organization focuses primarily on increasing awareness of the business benefits for business improvement. In the Take-off Stage, the focus is on developing working definitions and knowledge management strategies, which primarily focus on developing the structure, identifying required resources, and identifying barriers and risks. In the Expansion Stage, the focus is on providing organizational leadership. In this stage leadership behaviours are aligned with the knowledge management strategy and a structured approach to implementation and change management addresses barriers and risks. The Progressive Stage then focuses on improving the performance of knowledge management activities, which is characterized by an increased emphasis on monitoring and measuring performance. The final Stage, Sustainability, then focuses on the enabling the processes and approaches to become an integral part of how the organizational culture.

Although this represents a somewhat techno-centric focus, the assumptions underpinning these perspectives regard knowledge as something ‘out there’ ready to be captured, manipulated or leveraged (Gloet 2006). However, more recently a broader range of what constitutes knowledge is being purported that moves the definition of knowledge as an entity; i.e. information, to include a resource, a capacity or a process. Thus knowledge is seen as a knowledge building process; something that is created though social processes that has people and relationships at its centre, and are influenced by historical and linguistic contexts (Hosking 2011; Gloet 2006). This more contextualized definition of knowledge aligns more closely to the definition of knowledge management as proffered by Pillainia (2008). However, it also makes notions of knowledge much more intangible, which in turn adds to the complexity of doing business in a knowledge based economy. For example, according to Okada (2004) the knowledge and skills requirements for production are becoming more tacit, more firm specific and harder to transfer between organizations. To deal with this complexity, a model such as the STEPS Knowledge Management Maturity Road Map may be useful in guiding and informing the knowledge management process.

This is important, particularly as notions of knowledge management can be linked to sustainability, and offer up similar issues regarding knowledge creation and knowledge management. The recognition of the centrality of people, relationships and context within the knowledge management process has also been identified as central to the sustainability debate, as it is through human and social capital that sustainability development can be put into practice (Seow et al. 2006). Building collective organisational knowledge about sustainability and sustainability practices therefore is a knowledge management activity that can enable an organisation to create and meet its sustainability objectives. To what extent people in organizations are able to engage in knowledge building activities around sustainability often depends on the broad understanding about sustainability within the organisation. This means that existing knowledge about sustainability will most often result in the main initial driver for sustainability. Expanding the breadth and depth of understanding about sustainability will inevitably determine the scope of initiatives undertaken by an organisation (Cushman et al. 2012). Seow et al. (2006) concur but suggest that Corporate Social Responsibility (CSR) needs to be at the centre of the sustainability debate. This then opens up an avenue for addressing both the tangible (entity) and intangible (know-how and creativity) aspects of knowledge creation. The sustainability agenda of an organisation is ideally related to what it produces, how it goes about production and who (people) is involved in the processes of production. Seow et al. (2006) further propose that any corporate sustainability agenda should include all aspects and stakeholders of an organisation and includes the workers, the investors, the consumers and the wider community. An initiative ideally addresses working conditions, regulatory compliance, environmental impacts such as pollution, waste and rework and the establishments of environmentally sound projects as learning environments (Seow et al. 2006). Thus, according to Seow et al. (2006), key factors for a successful sustainability strategy revolve around establishing clear goals and alignment of knowledge management to business objectives and the creation of awareness and motivation. Providing leadership, resources, management support, staff and budget are also important, as is the availability of support of both IT and non-IT based tools. Lastly they suggest that any recognition of reform needs to address barriers, facilitate implementation and needs to be tied to clear performance measurers for evaluation.

Knowledge Building, Knowledge Management and Sustainability in India's Automotive Industry

In India, during the 1990's, its economy became increasingly integrated into the global economy. For the Indian automotive industry this meant that globalised production processes demanded not only a different skill mix but also vastly different employment practices. These demands required upgrades in India's skill base as both assemblers and suppliers, and it required that Indian firms acquire and form closer inter-firm linkages with both domestically as well as internationally. This demand changed the onus of workers' skill development from a requirement traditionally driven by market forces to a critical strategic issue driven by the need to participate and compete within in a global supply chain environment (Okada 2004). Thus an initial knowledge building process, as suggested by Gloet (2006) emerged, where in the Indian context these processes initially revolved around developing new relationships, new skills and new processes. This change influenced not only how Indian organisations developed their workers, but it also influenced their employment practices. The three skills areas requiring attention revolved around production skills, managerial skills and engineering skills (Okada 2004). India's rapid response to these requirements enabled it to catch up quickly and link into the industry's global value chain through licensing, collaborations and joint ventures with multinational corporations. The Indian automotive industry realized that a knowledge management approach would enable them to build on valuable existing skills as well as develop new skills. The Mahindra group was an early adopter of a knowledge management approach and was able to capitalize on this and develop a competitive advantage.

Sustainability

To provide context to the approach to sustainability taken by the Mahindra Group it is useful to understand the emergence of sustainability in Indian industry in general.

In 2007, 'sustainability' as a concept was largely an unknown phenomenon in India. This was not only the case within the Mahindra Group but also within the wider Indian corporate landscape. There were a few discussions taking place at the Government level and among Indian companies who had an off shore presence. A number of those had taken some initial steps towards understanding the issues associated with sustainability. More specifically, at the time, around 8–10 companies had begun to publish their Sustainability Reports based on the Global Reporting Initiative (GRI) framework.

This framework, launched in 1997, was a shared initiative between the United Nations Environmental Program (UNEP) and the United States non-governmental organisation, Coalition for Environmentally Responsible Economics (CERES). The GRI framework aims to enhance 'the rigour and utility of sustainability reporting'. As such, it provides guidelines with specific indicators to enable this reporting to take place in the following three specific categories; social, economic and environmental. Although not all of the indicators provided prove to be easy to evaluate, they do provide guidance as to what to consider at the operational and or project levels within an organisation (Labuschagne et al. 2005).

Furthermore, at the time, the fourth Intergovernmental Panel on Climate Change (IPCC) report was also being discussed at certain levels within Government, however, these discussions were also not part of the corporate agenda. This IPCC report focused attention on the need to integrate issues around climate change with sustainable development policies and the relationships between mitigation and adaptation strategies. This focus provided a closer link to the issue of Corporate Social Responsibility (CSR), a concepts which was well understood, prevalent among Indian corporations and a solid feature on the Indian corporate agenda.

However, the Indian definition of CSR differed from the definition used by developed economies. While in developed economies CSR refers to an overall responsibility of business to the environment as well as towards the communities in which they operate, in India it is restricted to the notion of 'Philanthropy'. This means that socially responsible corporations in India are committed to an accelerated and holistic growth of India that focuses on its people. Hence they have a legacy of philanthropic activities, which generally center on areas of health, education and livelihood. For the Indian companies who had a good track record in this philanthropic approach to CSR, the migration toward a more holistic 'Triple Bottom Line' approach and a review of their current approaches in order to enhance their sphere of responsibility to include sustainability proved to be straightforward.

Today, Indian Companies have begun to understand the importance of sustainable growth. The Government has announced a Nation Action Plan for Climate Change. Industry Bodies like the Confederation of Indian Industry (CII) and the Federation of Indian Chambers of Commerce and Industry (FICCI) are in continuous dialogue with the Government to review existing laws and propose new ones. This is done with the aim to create environments conducive for Indian Companies to adopt a path of sustainable growth. In addition, the financial regulator the Securities and Exchange Board of India (SEBI) and the Bombay Stock Exchange have started creating disclosure structures that would enable companies to compete in the carbon constrained economy of the future.

Mahindra & Mahindra

Background

Behind every successful Institution lies an interesting story that is often based on a strong legacy. This is certainly the case for the Mahindra Group. It can be said that an independent India and the establishment of the Mahindra Group largely coincided. In 1945, two enterprising brothers named J.C. Mahindra and K.C. Mahindra joined forces with Ghulam Mohammed and started Mahindra & Mohammed as a steel company in Mumbai. Two years later, upon the partition of the British Indian Empire, India won its independence from Great Britain and the independent state of Pakistan was created. As Ghulam Mohammed became Pakistan's first Finance Minister, the Mahindra brothers ignited the companies' enduring growth with their decision to manufacture 'Willys' jeeps under license in Mumbai. Thus the company became Mahindra & Mahindra Limited (M&M) here from referred to as the Mahindra Group.

Working under license was a common strategy employed by Indian family businesses, as the Indian policy framework of the time reserved whole industries for the state sector, placed restrictions on the scope of activities for the private sector, and subjected imports to quotas and licenses (Manikutty 2000). In addition to enabling The Mahindra Group to flourish in this difficult policy environment, the decision to manufacture Willy's jeeps under license was also based on the belief by the Mahindra brothers that new modes of transportation could be a key to India's prosperity. Therefore, one of their first goals was to build rugged, simple vehicles that would be capable of tackling the Indian terrain.

Prior to 1991, family business groups dominated the private sector in India. However, the Indian government's move towards market liberalization in 1991, which gradually reduced and dismantled government controls increased the level of direct foreign investment and the formation of joint ventures with international partners. As early pioneers of globalization, the Mahindra brothers collaborated with a wide range of international companies and before long, Mahindra & Mahindra's reach extended to steel, tractors, telecom, and more. Now, after more than six decades, Mahindra has grown from its humble local beginnings to a US $16.7 billion corporation that employs more than 180,000 people around the world. Today, the Group operates in 18 key industries that are at the foundation of every modern economy.

Mahindra & Mahindra Limited (M&M) is the flagship company of the Group, and the Holding Company for all of its wholly owned subsidiaries. Its core business is the manufacture of utility vehicles and tractors, a sector in which it has secured global leadership. However, since 1991, it has diversified into the IT, Financial Services and Leisure & Hospitality sectors. It conducts business in these sectors through separate entities with each entity having its independent Board and Management structure. In each of these sectors Mahindra has commanded a significant global presence. For example, it diversified and created businesses in the Aerospace, Construction Equipment, Consulting Services, Defense, Logistics, Energy, Real Estate, Retail, and the Two Wheelers sectors.

Organization and Governance Structure

The Mahindra group is organized around a federated structure. This means that each business is capable of charting its own future, whilst being able to leverage synergies and competencies across the entire group. Mahindra's leadership from its inception has endorsed this structure. As such, various structures and committees have been put in place in order to ensure that the diversity of expertise across the various businesses is leveraged and that the ethos and integrity of the Mahindra Core Values are maintained. These structures are as follows:

At the helm of the Mahindra Group is the Group Executive Board (GEB). This Board provides overall leadership to the Mahindra Group. Its membership consists of the chairman & managing director of the holding Company – M&M, presidents and chief executive officers of the various business as well as heads of certain key corporate functions. The role of the GEB is to provide strategic direction and enterprise leadership, facilitate synergistic and symbiotic relationships between the various businesses that make up the Mahindra Group and staying abreast of its overall direction. It does this by creating and driving a shared vision and value-system across the various business units and companies that make up the Mahindra Group; explore synergies between all businesses in order to unite the Group's goals and develop high level strategic plans.

To enable a consistent approach to the implementation of the GEB's goals and strategic directions, various special Committees and Councils have been formed. These Committees and Councils focus on specific issues or subjects. As such, they are charged with ensuring a whole of organisation approach within the Mahindra Group around matters that are critical to growth across all industries. These special Committees and Councils are:

- Group Sustainability Council
- Group Corporate Social Responsibility (CSR) Council
- Apex Talent Council
- Corporate Risk Committee
- Central HR Council.
- Sexual Harassment Committee
- Information Security Apex Committee
- Corporate Governance Cell
- Central Safety Council

Although these entities mandate and drive a strategic approach to specific issues throughout the Mahindra Group, each individual business within the Group is free to form their own committees to be able to approach and implement initiatives congruent with their own business context. This localized approach to implementation enables each individual business within the Group to approach matters from its own perspective whilst maintaining overall Mahindra Group alignment.

An additional key component to enabling strategic alignment, and leveraging synergies and key competencies across the Mahindra Group is its Corporate Centre. It ensures that the Group's companies have a common understanding of the Group's Core Purpose & Values and it assists in achieving consistency in operationalizing common tasks and services across the Group. As such, the Corporate Centre's departments include Corporate HR, the Office of Strategy Management, Corporate IT, Corporate Communications, Corporate Sustainability Cell, Corporate Legal, Corporate Finance, Corporate Affairs, Internal Audit and Corporate CSR.

Ethics, good governance and corporate social responsibility have been at the core of operations within the Mahindra Group. As, its Core Values reflect this stance and have been articulated as: Professionalism; Good Corporate Citizenship; Dignity of the Individual; Customer First; and, Quality Focus. In order to reinforce the adoption and application of these values, a new Brand Position 'Mahindra RISE' was embraced in 2010 across the Mahindra Group. 'Mahindra RISE' is built on 3 pillars to enable employees to actually live and embody the Core Values. The three pillars are articulated as: Accept No Limits; Alternative Thinking; and, Drive Positive Change. These Values and Brand Pillars form the basis of the redefined Core Purpose of the Group. It aims to extend Mahindra's reach to go beyond the organization by enabling employees to consciously touch the lives of external stakeholders as well. As a result, the RISE philosophy has already played a meaningful role in addressing issues arising from cultural, intergenerational and international diversity. In addition, it has contributed significantly to the global reinforcement of the Mahindra Brand, and it provides an impetus for sustainability within the Mahindra Group.

The Mahindra Approach to Sustainability

In order to effectively embed sustainability drivers within the Mahindra Group a knowledge building approach was deployed, which was closely aligned with the organization's values and ethos. Following is a description of the approach and an outline of activities and initiatives undertaken. It demonstrates how Mahindra employed a systematic approach to create awareness and understanding of the importance of integrating sustainability drivers at the strategic, managerial and operational level, using a common platform of "Alternative Thinking" which enabled a common understanding of issues and at the same time providing for the diversity of the business sectors, size and other variables. This enabled the different businesses within the group to identify areas in which they believed they could make a significant difference and allowed each company to decide its own plan forward.

Assessing Internal Strengths, Challenges and Stakeholder Expectations

Strengths

The first step taken by the Mahindra Group in its sustainability journey was to assess its internal strength and map, both internal as well as external challenges. This exercise made it clear that the organisation's strengths revolved around its Core

Values; its commitment to governance transparency & ethics; a commitment to social welfare initiatives which were closely aligned to the Indian National Inclusive Growth Agenda, and its existing culture of innovation.

A high employee commitment to the organization's core values at the Mahindra group was evident. It possessed well defined Codes of Conduct which were based on its Core Values for all staff, including Director and Senior Managers. These had been instrumental in guiding employee behaviour throughout the organisation and contributed to the creation of an organisational culture that was based on employee loyalty. The leadership of the GEB, the special Committees and Councils and the Corporate Centre played an important role to ensure that the highest standards of ethics and governance were maintained throughout the organisation. This strong commitment to transparency and ethics enabled and encouraged its businesses to go beyond compliance. For example, manufacturing units were encouraged to get Environmental Management Systems (EMS) and Health & Safety related certifications like ISO & OHSAS for all factories. Other business units like Mahindra Holidays had initiated the process for ECOTEL certification, demonstrates its commitment to environmentally friendly hospitality systems, processes and practices.

Another strength identified was the Mahindra Group's commitment to corporate citizenship (Mohan 2001), which was demonstrated through its commitment to contribute to social welfare aligned to the National inclusive growth agenda. While contribution towards social causes had always been high on the Mahindra Group's agenda a more structured approach was put in place in 2005. On its 60th anniversary it committed to set aside one percent of the Groups profits on an annual basis to be directed towards societal development every year. With the funds evenly distributed between the Mahindra Foundation and supporting employee volunteerism through its Employee Social Option Scheme (ESOP). The ESOP supports employees in their individual and team endeavours to contribute to the social welfare and development of their local communities. The commitment was further institutionalised by way of a governance structure consisting of a high profile Council to provide oversight and to enable a transparent distribution strategy which focused on national priorities such as health, education and environmental initiatives. The third strength which also clearly stood out was the existing 'innovation enabling' environment across the Group, deeply engrained in its existing culture.

Challenges

The initial assessment conducted also identified a number of challenges. The diversity in terms of levels of maturity of each of the businesses within the Mahindra Group proved to be a major challenge. Although the mature and profitable businesses within the group would be able to bring about the necessary changes, less mature businesses had clear financial and human resources constraints. In addition, the different realities faced by the businesses because of their industries also needed to be considered. For example, the competitor, customer and value chains of companies in the manufacturing, financial services, IT, and real estate are vastly different. Hence the ability to develop a shared understanding of the concept of sustainability and what this meant within the context of each business and across industry groups was identified as a major challenge. For an organisation like Mahindra that implements an effective knowledge management program could utilise the STEPS knowledge management initiative to attempt to collect and distribute sustainability focused knowledge within the organisation. While the federated structure of corporate governance within the Mahindra Group had its clear advantages, it also meant that every Chief Executive Officer and Chief Experience Officer had to be convinced of the merits of including sustainability levers in their business processes. In addition, the thousands of professionals with different priorities, domain knowledge and business challenges also warranted close consideration.

Furthermore, initial interactions with employees in various businesses and at different levels indicated that people had a vague understanding of the various aspects of sustainability. The relevance of 'climate change' to business was hardly ever discussed or debated. As such, initial discussions on the subject raised much cynicism across all levels of the organisation and its businesses. For example, businesses in the services sector felt that sustainability was relevant to only manufacturing businesses, and those in the manufacturing businesses felt that any change would mean a financial burden. Much of the initial resistance revolved around two specific areas; a perception of increased cost and a reluctance to having to change existing processes and practices. As it was felt that any change to adopting 'green' practices would be expensive, managers refrained from even considering a shift. In addition, companies which had developed and refined their respective frameworks for risk management, supplier selection, supplier audits, new project approvals etc., were reluctant to introduce any additional dimension to these frameworks.

Stakeholder Expectations

The most important development that changed the above perception was the radical changes in the external business landscape and stakeholder expectations. For example, it was noted that a trend was emerging whereby Institutional Investors, in addition to reporting under specific disclosure frameworks, were also keen to understand a business' resilience to climate change. B2B businesses were expected to disclose social and environmental performance for renewal of contracts. Changes in Government Regulations in terms of stricter Environmental protection laws, product labeling for energy efficiency in consumer products and compulsory disclosures for non- financial disclosures as a part of the Annual Report became a regular phenomenon. Following this, in 2013 the Bombay Stock Exchange launched a carbon Index with S&P called The S&P BSE-GREENEX Index, which includes the top 25 companies which are good in terms of Carbon Emissions, Free-Float Market Capitalization and Turnover. This step by the Country's premier Stock exchange brought about a kind of a confirmation that increased demands for non-financial disclosures from Governments, Regulators, Investors and other stakeholders will be the new norm.

Rather than trying to find excuses for not making disclosures, the approach adopted by the Group was to understand the kind of disclosures expected by the various stakeholders and respond to the same. Developing a deep understanding of the available disclosure frameworks actually helped in understanding the emerging issues and expectations of the external stakeholders, and kept the Group companies at the head of the learning curve.

Defining the Approach

In order for the Mahindra group to overcome these challenges, and capitalise on its strengths, it became clear that it needed to define a means that included the entire organization. Hence a top down as well as a bottom–up approach was adopted to ingrain the dimensions of sustainability within the ethos and culture of the organization. It was also decided to plan the process in stages (Fig. 8.1).

Within the Strategic Stage clear priorities were defined, to address the need for a 'Greening strategy, chart out a sustainability roadmap with priorities, understand the business case for sustainability in new projects and see how all of these can be leveraged for brand enhancement. Within the Managerial Stage a governance framework for sustainability was put in place, an internal auditing structure designed and mechanisms to measure and monitor sustainability performance formulated. In addition, the existing policy framework was reviewed through sustainability lenses. The focus for the Operational stage was to embed sustainability into existing processes from procurement to sales. Human Resources focused on developing sustainability awareness and including sustainability in key performance indicators, etc.

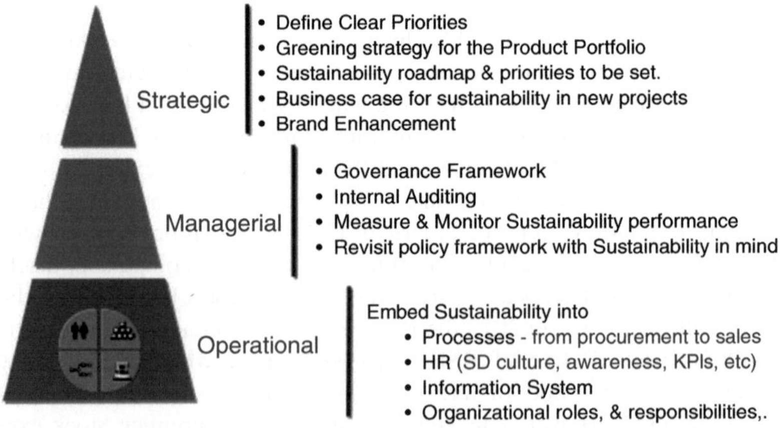

Fig. 8.1 The Mahindra group approach to sustainability. Mahindra group

Information Technology was used to support data collection, monitoring and review. In short, organisational system, roles and responsibilities were aligned to the sustainability requirements.

Following this internal approach, and after a review of available framework options for public disclosure, initially the GRI framework was adopted. The GRI framework enabled the Mahindra group to make structured external disclosures about its sustainability endeavours aligned to international standards. It was also decided at that time that because of the diverse nature of the businesses and the various stages of maturity of businesses within the Mahindra group, to make disclosure on an annual basis at a group level rather than at the level of individual business. It was therefore decided to start reporting for all large businesses in year one and add companies as they matured. In the first year, companies accounting for over 85% of the Group's turnover were scoped in the report.

The initial focus on encouraging business to disclose under the GRI framework in terms of environmental, social and economic parameters, evolved as the Mahindra Group developed a deeper understanding of the broader external contexts. Thus, Group companies were also increasingly encouraged to make disclosures under the Carbon Disclosure Project (CDP) and the Dow Jones Sustainability Index (DJSI).

Organizing for Sustainability

In order to provide direction and guide the implementation from a strategic level, the Mahindra group established a Sustainability Council (Fig. 8.2). This Council reports to the Board Committee for Corporate Social Responsibility, As such, the sustainability Council's tasks revolve around approving new initiatives and monitoring progress of the integration of sustainability into all parameters of the business and its operations. Members of the Sustainability Council are CEO's and CXO's of all companies as well as senior heads of functional areas such as Corporate Communications, Group HR, Group CSR, Quality etc. Representation of senior people at this strategic level is a key component to ensuring effective integration of Sustainability across all businesses and functions.

In turn, the Sustainability Council guides the work of the Group Sustainability Cell. The Cell's roles are to, (a) drive sustainability through awareness raising and knowledge building across the Mahindra Group; (b) provides practical support to individual businesses by assisting in developing the know-how to integrate sustainability in the strategic business processes and operations; and, (c) help businesses to identify risks and opportunities from climate change and suggest risk mitigation avenues.

Sustainability Champions are located at each plant or office and mostly represent members of the middle management ranks. They are identified by each business based on availability and type of business, as the Champions' role is to assist in contextualizing sustainability activities to each business. For example, most manufacturing businesses' Sustainability Champions are the Energy or Utilities Heads which operate at each location. However, for the Services Companies, the Champions

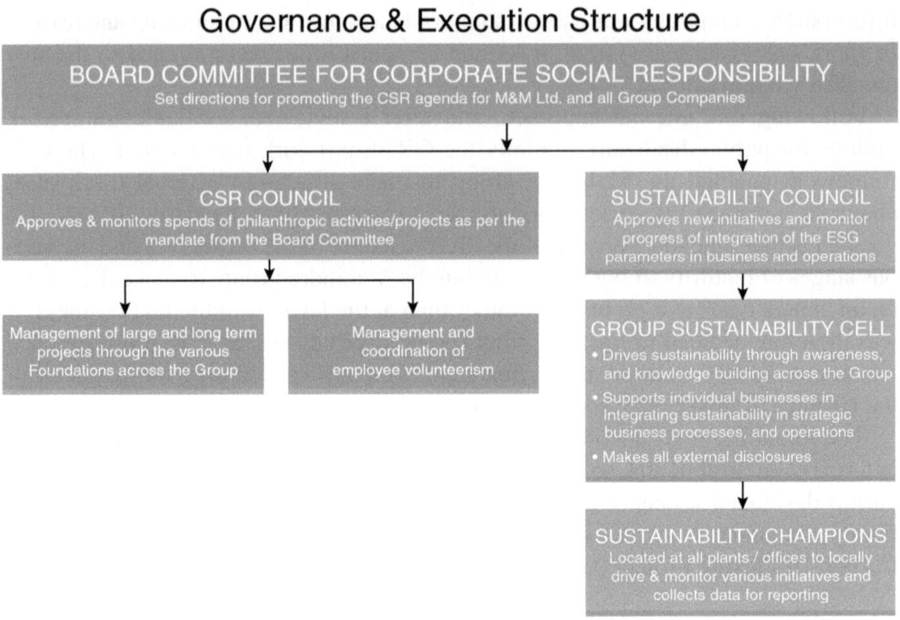

Fig. 8.2 Governance & execution structure sustainability. Mahindra group

are mostly selected out of the senior HR ranks. The Champions are supported by a team of three of four additional people who are mainly involved in collection of data, and reporting related work. None of the Champions have 'Sustainability' as a full time responsibility, but manage it as a part of their existing portfolio.

Raising Awareness

As a lack of awareness about the dimensions of sustainability in general was identified as one of the major challenges, an awareness raising strategy was defined and planned. A major component of the strategy revolved around branding the concept of sustainability from a Mahindra Group perspective. As earlier internal research within the organization had revealed, sustainability meant different things to different people. Ensuring a common understanding across the organization had to be ensured. A series of interviews, surveys and interactions with employees from different businesses and domains with the Group, coined the term 'Alternative Thinking' as integral to overcoming issues of sustainability. The premise on which this was based was to suggest that alternative thinking precedes innovation and change in whatever one does. The thought was inspired by the famous quote of Albert Einstein "We cannot solve a problem by using the same kind of thinking which created it." In addition to promoting the need for changes in the way in which business was conventionally approached, the idea of 'Alternative Thinking' also aligned well with the culture of innovation which was historically promoted across the Group. The 'Alternative Thinking' concept continues to be used in all internal communications relating to Sustainability. In addition, although the term was coined in 2007–08, the term was included as one of the three key pillars of the new brand positioning 'Mahindra RISE' in 2010. According to employees at the time, they believed that 'Alternative Thinking' represented what the Mahindra Group stood for and how they would like to be perceived by the external world.

To create general awareness about the various aspects of sustainability was also perceived as a critical task. This was done through various media such as posters, banners, quiz competitions, pledges, messages on the corporate intranet, E-zines, in-house magazines etc. The messages contained in these communications revolved around the following three major aspects; Climate Change & Global warming; the need for behavioural change by employees; and, the need for change in attitudes towards the environment. Banners, poster as and e-mailers were created under the brand banner of 'Alternative Thinking'. For example, some of these messages read 'Collaborate with Nature and disseminate climate change best practices to a wider audience'– give every product and service a green edge; Start an Alternative Savings Plan – Save water today – avert a thirsty tomorrow; and others. The Sustainability Champions at each location were instrumental and played a critical role in this awareness raising. To date, specialized awareness programs around sustainability still continue in terms of organized events like 'Environment Week' or 'Water Day' etc. These ongoing initiatives have helped employees across all levels of the organization to understand the challenges that need to be addressed in their work as well as in their personal lives.

Creating Management Buy-in

In order to enable managers to integrate change into their strategic planning and their daily work routines, specialized programs with a focus on articulating the business case for sustainability were prepared. Tailor made presentations were designed for the senior leadership teams of each business as well as the heads of various specialized functions e.g. Accountants, Human Resources, IT, Merger & Acquisition teams, Plant & Function heads, Sourcing & Supply Chain, Special Projects teams, Strategy & Risk Teams etc. Each presentation was backed by examples of best practice examples and offered benefits & risks experienced by other companies in similar industries. A number of specific areas of relevance proved most impactful in convincing managers that considering sustainability made good business sense. These were potential regulatory changes focusing on a future carbon-constrained economy and changing disclosure expectations from foreign institutional investors.

Specifically, as changes to address a carbon-constrained economy were evolving, both in India as well as in international markets, it was evident that it was increasingly important for business heads and Chief Financial Officers to align business strategies to such changes. In addition, the probability of such future change in regulation, the risks of not being pro-active to such change and the opportunities presented by taking the lead were also explained. For example, the Clean Development Mechanism, which was most relevant at the time, was identified as one of the flexibility mechanisms defined by the Kyoto Protocol (IPCC 2007) to provide means for emission reduction projects. These could then be traded in emission trading schemes, such as the Perform Achieve Trade (PAT) scheme in India designed to reward carbon efficiency and penalize inefficiency.

In addition, changing disclosure expectations from foreign institutional investors like Pension Funds, through Carbon Disclosure Projects (CDP) and the emergence of Carbon Indexes like the Dow Jones Sustainability Index (DJSI) also identified clear risks and opportunities. On a more practical note, for Energy Managers, specifically focused training programs were conducted through the Mahindra Institute of Quality. These 3-day intensive programs enabled participants to develop the required insights and skills. At the conclusion of the 3 days, participants made a special commitment to take on a project at their respective factory locations. Following from there, the Group Sustainability Cell follows up by monitoring progress and by providing support such as identifying external agencies that can provide additional expertise, undertaking preliminary energy audits, identifying new technologies, etc. Since 2012, these programs have been extended to provide suppliers with the opportunity to participate.

In order to counter negative mindsets, plant managers were provided with incentives. This not only proved an effective strategy, but also led to early positive results. For example, The Group Sustainability Cell convinced Plant managers to allow them to carry out 'walk through' audits for energy and water use. These audits were carried out at no charge to the Plant. However, in exchange managers were to make a commitment that if a project could be identified, the project would be funded out of the savings incurred by not having to have paid an external agency for an audit. In addition, any savings as a result from undertaking the project were to be allocated to new resource saving projects at the Plant. Through these audits, potential projects were identified and classified according to three specific categories, (1) projects with zero to marginal investments; (2) projects with a less than 2-year payback; and, (3) long term, high investment projects. Achieving buy-in into the first two categories proved easy. Once the Plant heads and managers realized the cost benefits, word quickly spread and drove positive change across the organization. Projects focusing on reducing energy, water and packaging waste quickly took hold.

To facilitate the collection of results and other relevant information critical to the reporting of successes, Capacity Building workshops (CBWs) were arranged to equip the Champions and the other support staff to understand the requirement under the reporting framework. This data collection process was backed by an IT enabled program, which was designed and developed in-house. Continuous upgrades of this IT program are now an annual process undertaken by the sustainability cell. Close to 200 employees are taken through the CBWs on an annual basis. Other initiatives currently undertaken by the Group Sustainability Cell is the maintenance of a best practices bank which is can be accessed by function managers via the Intranet for the function managers to access. In addition, sustainability related became part of Induction Training, which enables new recruits to understand the aspects of sustainability. Also, regular needs based training is undertaken at the request of individual businesses.

The Mahindra Sustainability Road Map

In order to chart a course for sustainability that can be measured and monitored effectively, a Mahindra Group Sustainability Road map was designed. The road map revolved around seven specific commitments, which were relevant to everyone. In the addition to providing initial direction, roadmap also outlined small targets designed to ensure buy-in from the operational level. The seven commitments were identified as: reducing energy consumption; reducing resource consumption; reducing greenhouse gasses & CO2 emissions; green IT & green procurement; earning green certifications; spreading sustainability awareness to stakeholders; and, increasing employee engagement through ESOPs. Based on these commitments managers were encouraged to choose projects with clear cost benefits. Positive results achieved at one location encouraged others to follow suit and good examples were shared and widely acknowledged. This led to a snowballing effect, which enabled in many cases saw 5-year targets surpassed in 3 years in four out of the seven commitments. These results gave confidence to smaller teams to take on projects within their areas, and it countered the myth that all 'green' projects need high capital investments.

While the diversity of issues relevant to driving sustainability within each business was recognised and encouraged, the initiatives undertaken were analysed and categorised in order to identify common patterns. This resulted in a better understanding of the strategic impact of initiatives. As such, it was determined that most sustainability projects within the Mahindra Group revolved within the following five areas: Eco-efficiency, Green Supply Chain, Responsible Product Development, Employee Care and Development, and Community Development. The Sustainability Road Map proved to be the most effective tool in driving sustainability within all businesses. Based on what was learned from implementing the first Road Map, further developments in the area of sustainability a second Sustainability Road Map is being developed.

In 2012, the Mahindra Group Sustainability Awards were instituted. These awards recognize success, and further encourage and celebrate innovative and transformational sustainability efforts within the Mahindra Group. In addition, the Awards are a high profile means by which to transforming the concept of sustainability from theory to practice across the Group. Also, it aims to motivate businesses to accelerate their progress with respect to sustainability issues and providing further understanding of the strategic context of sustainability related risks and opportunities. It contributes to the ability to measure intra & inter sector sustainability best practices and a scaling up of innovative concepts. Finally, it contributes to assessing the effectiveness of the implemented initiatives and contributes to achieving sustainability excellence within the Mahindra Group.

Achievements and Outcomes to Date

To date, the Mahindra Group has been able to position a focus on sustainability as an integral part of how it conducts its business. From an internal perspective, it has derived significant financial benefits from its reductions in energy consumption of almost 20% and water consumption of 28% in 5 years. Climate change and environmental, social and government factors are now considered in strategy, risk management and new product development etc. as a matter of course. Managers see a clear benefit in terms of cost efficiencies and there are improved trends in the consumption of natural resources. In addition, funding for projects and investments in new businesses that enable eco-efficiencies are encouraged. For example, the Mahindra Group has invested $100 million in a Research & Development facility in Chennai. It has investments in electric cars, solar power generation and micro irrigation systems. It has committed to planting a million trees each year and is continuing its focus on water efficiency projects.

From an external perspective, the Mahindra Group has seen a clear and positive effect on its brand. For example, three of its listed Companies i.e. Mahindra & Mahindra Limited, Tech Mahindra and Mahindra & Mahindra Financial Services Limited are now part of the DJSI (Emerging Markets) as well as listed in the Carbon Disclosure Leadership (India) Index. Furthermore, our companies are rated by financial analysts based on information in public domain. Most recently, (in September 2014) Mahindra & Mahindra Limited Ranked 33 in the list of top 100 companies across ten key Asian economies, by Channel NewsAsia Sustainability Ranking – Sustainalytics.

Its focus on voluntary disclosure in the public domain, has led to a keen interest in its businesses by institutional investors. In addition, the RISE philosophy has caught the imagination of many people. Not only has it ignited internal business transformation among employees, it has also inspired people across all walks of life. In particular, the 'Spark the Rise' campaign launched in 2012, showed an overwhelming response from over 267,648 people who were eager to connect and act. It recorded 1470 projects wherein people made commitments to take responsibility for a better future of their communities.

External Collaborations and Partnerships

In order to stay abreast of developments in sustainability practices nationally and internationally, and to take on a leadership role, the Mahindra Group has developed a number of external collaborations and partnerships. It is a member of the Business

Council for Sustainable Development, an initiative of the Environment Research Institute of India. It represents an ongoing collaborative effort between researchers of the Environment Research Institute and industry leaders to address issues in Indian industry. As such, it provides and independent platform for corporate leaders to address issues related to sustainable development and promote leadership in environmental management, social responsibility, and economic performance.

The Mahindra Group is an active participant in Sustainability committees of leading Industry bodies like the Confederation of Indian Industry (CII) and the Bombay Chamber of Commerce & Industry (BCCI). Through lobbying and advocacy efforts, it promotes the sustainability agenda with regulators and Government Bodies and among other Indian corporations.

The Mahindra Group is also a member of the United Nations Global Compact (UNGC) and actively promotes the integration of the Ten Principles of UNGC in Business by leading the Western Region of the Global Compact Network India.

Besides these collaborations the Group also finds value in participating in Working Groups formed for understanding and developing tools and business cases to be able to address the new and emerging expectations of the external stakeholders. The Mahindra Group is associated with

- The World Resources Institute (WRI) for developing a tool for calculating GHG emissions from Supply chains,
- The World Business Council for Sustainable Development (WBCSD) for creating the India Water Tool.
- CII for promoting the Business & Bio-diversity program in India.
- CII to understand and design the Integrated Reporting framework.

These collaborations enhance the Mahindra Group's leadership position/profile as a thought leader and a responsible organization in the area of sustainability.

Conclusion

The Mahindra approach to sustainability is a clear example of a knowledge building/knowledge management exercise following a methodical approach (Okada 2004; Gloet 2006; Seow et al. 2006).

It was understood at Mahindra that since the basic concept of 'Social Responsibility', which in the Indian context meant philanthropy was well ingrained in the DNA of the Mahindra group, it would help in incorporating the angle of 'Business Responsibility' as well. Corporate Governance, accountability & Transparency was also inherent in the Group's ethos. Once the results of integrating Sustainability drivers in business were seen, all the three aspects of Business responsibility i.e. Environmental, Social & Governance were merged under one platform. As of 2014, Mahindra has linked its CSR, Sustainability and Corporate Governance under a single branding umbrella of 'RISE for Good'. Through this, it can continue to link its sustainability initiatives into the wider community in which it operates.

When reviewing the Mahindra approach in light of the STEPS Maturity roadmap introduced by Seow et al. (2006), it becomes clear that many of the early activities undertaken by the Mahindra group resemble these suggested stages. STEPS reflects the various stages in which the knowledge management process matures. As it can be used to both analyze the progress made in the maturity of knowledge management processes within an organisation and guide the work, it thus enables an organization to identify at which stage it is operating, and what to do next. The case study in this chapter clearly shows that the approach taken by the Mahindra group reflects these stages. However, the Mahindra Group was wise to contextualize this to and Indian context. As such, it focused on increasing awareness and conceptual understanding of sustainability in a way that would matter to its workforce, by, for example, linking it to Indian conceptual understandings of CSR. In addition, it focused in particular on enabling Group businesses to contextualize sustainability as relevant to their individual business realities. Also, although it developed an overall governance structure that provided guidance and support, it allowed businesses to go about it at their own pace. Rather, it used formal and informal communication pipelines, networks and peer pressure to motivate and share ideas. Thus, it has transformed its businesses and sustainability is now considered as an integral part of how business is conducted today.

Sustainability is a journey and not a destination. Mahindra has thus far seen the benefits of taking this path in terms of business benefits, being ahead of the industry and gaining the advantage of setting the benchmark for other Indian businesses. It is ahead of many worldwide in terms of ratings on international indexes, and is reaping the benefits from its Brand perception and recognition. Its challenge is now to keep pace with the changes in the external business environment as well as maintaining its leadership position as others are following its lead and are also adopting sustainable practices. Six years into its journey, issues like resource efficiencies, green house gas reductions, Occupational Health & Safety, extending awareness to the supply chains etc. have become common practice.

For its second phase, the Mahindra Group is planning for radical shifts in its current agenda, which fall under the four specific areas. The first area is to sharpen its focus on issues that are material to its businesses as well as key stakeholders. The second area will focus on a deeper integration and operationalization of sustainability drivers within its sourcing and supply chains. It plans to intensify audits and will expect disclosures on resource consumptions and efficiencies from its sub-tiers. It is believed that this will help provide a better life cycle assessment of its products and services. Its third area of attention will focus on aspects of Ethics & Governance, Business & Human Rights and Anti-corruption, by establishing new expectations and engaging with external stakeholders like financial analysts, investors and regulations in all of the geographic regions where the Group operates. Its final area of attention will be on further strengthening the knowledge pool within the organization and continue to promote the initiatives that promote innovation and thought leadership within the Mahindra Group.

The Mahindra group's vision is to rank among the top 50 most admired brands by 2021. It believes that its focus on sustainability will be one of the most potent drivers towards achieving this vision.

References

Cushman, M., W. Venters, T. Cornford, and N. Mitev. 2012. Paper presented at British academy of management conference: Fast-tracking performance through partnerships, 9–11 September 2002, London, UK.

Gloet, M. 2006. Knowledge management and the links to HRM. *Management Research News* 29(7): 402–413.

Gold, A.H., A. Malhotra, and A.H. Segars. 2001. Knowledge management: An organizational capabilities perspective. *Journal of Management Information Systems* 18(1): 185–214.

Hosking, D.M. 2011. Telling tales of relations: Appreciating relational constructionism. *Organization Studies* 32(1): 47–65.

IPCC. 2007. *Climate change 2007 – Impacts, adaptation and vulnerability contribution of working group II to the fourth assessment report of the IPCC* (978 0521 70597-4 Paperback).

Labuschagne, C., A.C. Brent, and R.P.G. van Erck. 2005. Assessing the sustainability performances of industries. *Journal of Cleaner Production* 13: 373–385.

Lengnick-Hall, M., and C. Lengnick-Hall. 2003. *Human resources management in the knowledge economy*. San Francisco: Berrett-Koehler.

Manikutty, S. 2000. Family business groups in India: A resource-based view of the emerging trends. *Family Business Review* 13: 279–292.

Mohan, A. 2001. Corporate citizenship. *Journal of Corporate Citizenship* 2: 107–117.

Okada, A. 2004. Skills development and interfirm learning linkages under globalization: Lessons from the Indian automobile industry. *World Development* 32(7): 1265–1288.

Pillania, R.K. 2008. Innovations and knowledge management in emerging markets. *Knowledge and Process Management* 15(3): 184–185.

Seow, C., R. Hillary, H.S. Robinson, C.J. Anumba, P.M. Carrillo, and A.M. Al-Ghassani. 2006. STEPS: A knowledge management maturity roadmap for corporate sustainability. *Business Process Management Journal* 12(6): 793–808.

The Mahindra group's vision is to rank among the top 50 most admired brands by 2021. It believes that its focus on sustainability will be one of the most potent drivers towards achieving this vision.

References

[illegible]

MIX
Papier aus verantwortungsvollen Quellen
Paper from responsible sources
FSC® C105338

If you have any concerns about our products,
you can contact us on
ProductSafety@springernature.com

In case Publisher is established outside the EU,
the EU authorized representative is:
Springer Nature Customer Service Center GmbH
Europaplatz 3, 69115 Heidelberg, Germany

Printed by Libri Plureos GmbH
in Hamburg, Germany